Software Engineering for Automotive Systems

Software Engineering for Automotive Systems

Principles and Applications

Edited by

P. Sivakumar, B. Vinoth Kumar, and R. S. Sandhya Devi

CRC Press
Taylor & Francis Group
Boca Raton London New York

CRC Press is an imprint of the
Taylor & Francis Group, an **informa** business

First edition published 2022
by CRC Press
6000 Broken Sound Parkway NW, Suite 300, Boca Raton, FL 33487-2742

and by CRC Press
2 Park Square, Milton Park, Abingdon, Oxon, OX14 4RN

© 2022 selection and editorial matter, P. Sivakumar, B. Vinoth Kumar, and R. S. Sandhya Devi; individual chapters, the contributors

First edition published by CRC Press 2022

CRC Press is an imprint of Taylor & Francis Group, LLC

ISBN: 978-0-367-64785-8 (hbk)
ISBN: 978-1-032-21759-8 (pbk)
ISBN: 978-1-003-26990-8 (ebk)

DOI: 10.1201/9781003269908

Typeset in Times
by KnowledgeWorks Global Ltd.

Contents

Preface

SOFTWARE ENGINEERING FOR AUTOMOTIVE SYSTEMS: PRINCIPLES AND APPLICATIONS

This book explores the concept of principles and applications of software engineering in automotive systems for both beginners and advanced software engineers. This book presents the state of the art, challenges, and future trends in automotive software engineering.

The amount of automotive software in today's cars is inevitable in the current scenario, and working in the EV industry and on futuristic flying cars seems to continue in the years to come. The Indian automobile sector is predicted to have a prominent impact on the global economy and to be a promising employment domain for young minds, directly or indirectly, in the forthcoming years.

This book mainly targets this group or audience such as professionals working with automotive software and software engineering students who need to understand the specifics of automotive software. The book covers all the basic essential aspects of the automotive software field. As the first part, it introduces the topic related to AUTOSAR, communication protocols used in the automotive domain, and then it addresses the bootloader design of automotive software development, software architecture for autonomous trouble code diagnostics, automotive grade Linux for connected cars, edge computing for automotive applications, and nanosensors for future automotive applications. This book serves as a reliable, complete, and well-documented source of information on automotive software systems.

The chapters of this book will give a wide range of analysis on principles and applications of software engineering for automotive systems. A brief introduction to each chapter is as follows:

- Chapter 1 discusses the role of AUTOSAR in automotive software trends and the difference between traditional and adaptive AUTOSAR.
- Chapter 2 depicts the need for a common protocol in automotive communication systems and outlines the protocols used in automotive systems.
- Chapter 3 illustrates the design of bootloader software for an ADAS system and also discusses the security challenges faced during the updating of software and applications in the ADAS systems.
- Chapter 4 presents the notion of design requirements generally and describes the categories of requirements used when designing automotive software systems by emphasizing the identification of problems or challenges faced in automotive requirement engineering with reference to communication and organization structure.
- Chapter 5 proposes an on-board diagnostic system that has software architecture to solve the issues and alert the users with the help of troubleshooting codes.

- Chapter 6 discusses the open-source architecture for connected cars using automotive grade Linux to allow the accelerated creation of new technology and applications.
- Chapter 7 discusses the application of edge computing in the automotive domain. Usage of edge computing increases the computational capability of the nodes, enhances the application delivery time, and reduces the scope of congestion.
- Chapter 8 deals with nanomaterials and nanostructures used for the development of nanosensors for automotive applications.

We are grateful to the authors and reviewers for their excellent contributions in making this book possible.

We are grateful to CRC Press and Taylor and Francis Group, especially to Gauravjeet Singh Reen (Senior Commissioning Editor) for the excellent collaboration and support.

This edited book covers the theory and case studies of software engineering for automotive systems. We hope the chapters presented will inspire researchers and practitioners from academia and industry to spur further advances in the field.

Dr P. Sivakumar
Dr B. Vinoth Kumar
Ms R. S. Sandhya Devi
July 2021

Editor Biographies

P. Sivakumar received his B.E. degree in Electrical and Electronics with I class in 2006 from Anna University. He completed his M.E. degree in Embedded System Technologies with I class in 2009 from Anna University Coimbatore. He completed his Ph.D. in Electrical Engineering with a specialization in Automotive Embedded Software in 2018 from Anna University, PSG College of Technology. His research interests include embedded systems, model-based design, model-based testing of automotive software, automotive software development, fog and edge computing. He serves as a Guest Editor/ Reviewer of journals with leading publishers such as Inderscience and Springer.

B. Vinoth Kumar is working as an Associate Professor with 17 years of experience in the Department of Information Technology at PSG College of Technology. His research interests include computational intelligence, digital image processing, and embedded systems. He is the author of more than 40 papers in refereed journals and international conferences. He has edited four books with reputed publishers such as Springer and CRC Press. He serves as a Guest Editor/Reviewer of many journals with leading publishers such as Inderscience and Springer.

R. S. Sandhya Devi received her B.E. (Electronics and Communication Engineering) from Avinashilingam University, Coimbatore, Tamil Nadu, India. She completed her M.E. (Embedded Systems) at Anna University of Technology, Coimbatore, Tamil Nadu, India. She is an Assistant Professor at Kumaraguru College of Technology. Her areas of interest include embedded system design and ARM processors. She published her project in a journal and international conference.

1 Role of AUTOSAR in Automotive Software Trends

P. Sivakumar and A. Pavithra
Department of EEE, PSG College of Technology,
Coimbatore, India

S. K. Somasundarum
Department of IT, PSG College of Technology,
Coimbatore, India

P. K. Somanathan
ZF Friedrichshafen AG, Solihull, England,
United Kingdom

A. Manimuthu
Cybersecurity Centre @ NTU (CYSREN), Nanyang
Technological University, Singapore

CONTENTS

DOI: 10.1201/9781003269908-1

1

1.1 INTRODUCTION

Automotive open system architecture (AUTOSAR) is a plug-and-play platform architecture that allows vehicle original equipment manufacturers (OEMs) and Tier 1 provider to increase electronic control unit (ECU) programming efficiency, reduce development costs, and prevent re-advancement of identical ECU software program elements for the same vehicle applications. It is a new and evolving way of describing a layered programming system. AUTOSAR aims to enhance complexity management of built-in E/E architectures by allowing OEMs and vendors to reuse and swap software modules (Devi, R. S., Sivakumar, P., & Balaji, R. 2018). AUTOSAR aims to standardize the structure of ECU. AUTOSAR sets the stage for cutting-edge digital systems to improve efficiency, safety, and protection, among other things.

Automotive attributes such as independent driving, vehicle-to-everything (V2X) availability, over-the-air (OTA) refreshes, prescient support, and a wide range of current focuses are fundamentally founded on in-vehicle programming capacities. Each ECU must be functional in order for each of these features to operate flawlessly and to take into account continuous in-vehicle capabilities. To help complicated vehicular tasks, cutting-edge top-of-the-line engines have up to a hundred ECUs that communicate with each other using Ethernet, Controller Area Networking (CAN), FlexRay, CAN Flexible Data-Rate (FD), and other communication protocols (Elbahnihy, A., Safar, M., & El-Kharashi, M. W. 2020). OEMs previously used ECU software applications that run on a variety of platforms. There was no fashionable software program design being used by level 1 or Tier 1 providers and its merchants to coordinate ECU programming with OEMs previously. As a result, every time an OEM wanted to move to a different grade 1 or Tier 1 provider, or vice versa, the transition used to be extremely difficult. As a result, the new provider's decision to run an ongoing test in the middle via its formation presence period was almost irrational.

Tier 1 vehicle providers, semiconductor manufacturers, software providers, instrument providers, and others formed the AUTOSAR consortium in 2003 to enhance OEM-Tier 1 provider cooperation, improve ECU programming satisfaction, and reduce expenses and time (Honekamp, U. 2009). AUTOSAR is a flexible and standardized vehicle-programming program that aids in the optimization of interfaces among utility programming and vital vehicular features as well as the establishment of continuous ECU software program design throughout all AUTOSAR members. AUTOSAR can provide members with intrinsic benefits that enable them to monitor highly complicated E/E in-vehicle constraints, such as easy integration and exchange of features in complicated ECU people groups, and oversight of the complete item lifecycle (Kim, J. W., Lee, K. J., & Ahn, H. S. 2015; Sandhya, D. R., Sivakumar, P., & Balaji, R. 2019). AUTOSAR has given four chief arrivals of the normalized vehicle software structure and one arrival of acceptance tests about its classical platform since 2003. AUTOSAR's constructs are divided into three stages:

Stage I (2004–2006) Basic enhancements to general environment
Stage II (2007–2009) Standardization of structure and procedure
Stage III (2010–2013) Rehabilitation and improvements as required

In 2013, the AUTOSAR group launched an uninterrupted working mode for its traditional model in order to maintain the well-known while providing selected upgrades. The fundamental advancement exercises had been posted along these lines in some kind of a shared arrival of AUTOSAR classic, foundation, and adaptive, with release 18.10 in October 2018.The automotive software program is a collection of instructible data development that is used to carry out computerized in-vehicle operations. In-vehicle microcontroller's software is often referred to as automotive software (Sivakumar, P., Vinod, B., Devi, R. S. 2016a). Telemetry, multimedia, power train, body manipulation and convenience, connectivity, and enhanced driver assistance systems (EDAS) are examples of computerized in-vehicle implementations. Automotive software refers to the hardware and interfaces that run in-vehicle-integrated functions walkthrough on a machine (Mahmud, N. et al 2018; Sivakumar, P., Vinod, B., Devi, R. 2016b).

1.1.1 KEY DRIVERS

- Vehicles with EDAS are becoming more popular.
- The use of connected car services is becoming more widespread.
- Involvement of cutting-edge technology for modern user interfaces.

1.1.2 Key Restraint

- Inability to build software platforms using standard protocols
- Inadequately linked infrastructure
- Vehicle device's automotive software troubleshooting and repair

1.2 CLASSIC PLATFORM

The AUTOSAR Exemplary Stage design perceives the most significant degree of interpretation among three programming frameworks that operate on a microcontroller (Arts, T. et al 2015):

1. Application software component (ASW)
2. Runtime environment (RTE)
3. Basic software (BSW)
   - The ASW frame work is mostly hardware-agnostic.
   - Correspondence between programming sections and permission to BSW by methods for RTE.
   - The RTE covers the whole device interface.
   - BSW is made up of three layers, each with its own set of dynamic drivers:
     - Administrations, ECU reflection, and microcontroller consideration.
     - Administrations are categorized other than interested in valuable social affairs addressing the establishment for system, memory, and correspondence organizations.

1.3 ADAPTIVE PLATFORM

AUTOSAR-based adaptive platform implements the AUTOSAR runtime for adaptive applications (ARA) (Fürst, S., & Bechter, M. 2016). Services and Application Programming Interface (APIs) are the two types of connections required. The framework is made up of functional clusters that are integrated into services as well as the adaptive AUTOSAR foundation.

Functional clusters…

- Compile the functions of the adaptive framework,
- Organize the criteria specifications into groups, and
- Illustrate the software platform's actions from a network and an application standpoint.
- The adaptive platform's final SW specification of the programming model allows for a minimum of one service per (virtual) unit, although services can be spread through the vehicle arrangement.

The environment AUTOSAR API for agile technology platforms and adaptive users at runtime is superior to the AUTOSAR classic platform.

1.4 MARKET FOR AUTOMOTIVE SOFTWARE AND CORE TECHNOLOGIES

1.4.1 CONSOLIDATION OF ECUs

The primary goal of ECU consolidation is to reduce the number of pieces of hardware within the vehicle. This is usually accomplished by virtualizing hardware with software through a guest or host operating system (OS) architecture (Ryu, H., Jnag, S. Y., & Lee, W. J. 2013). Many vehicles have a mix of safety-significant and non-safety-significant structures. Unified cockpits are an example of this, in which the safety-significant (able to run on RTOS) and the IVI device (able to run on Linux OS) share the similar hardware, which includes storage unit.

1.4.2 UPDATES RECEIVED OTA

Automotive OTA improvements are a method of communicating clinical and functional info from a virtual server to interior structures including parameters of the vehicle in order to upgrade a variety of car applications such as ECU algorithms and multimedia without having to visit a workshop, vendor warehouse, or vendor location. Firmware OTA (FOTA) and Software application OTA (SOTA) updates are two forms of OTA improvements.

1.4.3 ROLE OF ARTIFICIAL INTELLIGENCE (AI) IN AUTOMOTIVE

In automobiles, electronics structures and strategies based on AI can be used. The machine's software is in charge of performing difficult operations and delivering learning capabilities. Many trends have emerged in the field of AI software applications, frameworks, and associated developer tools in recent years. Companies like Microsoft, Alphabet, and Intel are among the pioneers in AI software development. The majority of the companies involved in the AI logistics market's software application segment are based in North America.

1.4.4 MISSION OF AUTOSAR

The AUTOSAR association was established for the purpose of standardizing a typical device, critical machine features, and useful interfaces. This allows advancement allies to organize, exchange, reuse, and pass functions within an automobile group, dramatically increasing their production performance. AUTOSAR moves the change in perspective from just an ECU-based to a restriction contraption configuration effort in vehicle programming improvement and makes the organization of the constantly changing programming and E/E multifaceted nature for production and financial aspects.

1.4.5   PURPOSE OF AUTOSAR

AUTOSAR is a global development collaboration of vehicle-interested parties that was established in 2003. Its aim is to develop and implement a flexible and modular software program framework in automobile computer-assisted control systems.

1.4.6   METHODOLOGY

The system configuration description includes all framework records as well as information shared among various ECUs (for example, meaning of transport or BUS signals). The ECU-specific statics contain the statics from the configuration management description that are needed for an exact ECU (for instance, those alarms in which a remarkable ECU is approaching). The ECU setup summary is a file that contains all of the essential computer program setup information for a specific ECU.

1.5   AUTOSAR'S HIGHLIGHTS

1.5.1   ENHANCE SECURITY AND SAFETY

- Standard fault and threat scenarios are supported.
- Extend the testing and verification process.
- Streamline procedures.

1.5.2   CONNECTIVITY SHOULD BE IMPROVED

- Growing the number of mainstream cloud services.
- Consider the AUTOSAR AppStore.
- Connectivity to zone smartphones and control devices should be allowed.

1.5.3   DEVELOP SERVICE LIFE ENHANCEMENTS THAT ARE ADAPTABLE

- Enhance modularity
- Specify the cluster interfaces
- Ensure that funds are available to complete the device definition

1.6   AUTOSAR ARCHITECTURE

In the classic platform shown in Figure 1.1, the layered architecture supports

- Abstraction of hardware
- Runnable and job scheduling (OS) AUTOSAR
- Services for diagnosis and therapeutics
- Services for security and safety

AUTOSAR is open framework engineering for vehicle programming advancement and gives guidelines to creating normal vehicle programming applications. It is a

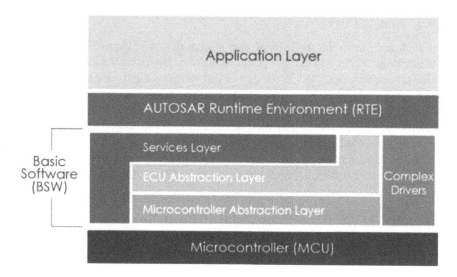

FIGURE 1.1 AUTOSAR architecture

growing and evolving standard that defines a layered architecture for software. The AUTOSAR tier, for example, is separated into three layers and operates upon the microcontroller. Let's take a closer look at these:

1.6.1 BSW Architecture

The AUTOSAR simple software structure is made up of many software modules organized in layers and shared by all AUTOSAR ECUs. It is therefore the company that developed BSW would share it with other companies working on engines, gearboxes, and other components. The BSW program architecture of AUTOSAR is made up of three layers:

1.6.1.1 Microcontroller Abstraction Layer (Mcal)

MCAL implements the special microcontroller interface and is also referred to as the hardware abstraction layer. MCAL offers drivers such as gadget drivers, diagnostic drivers, storage drivers, conversation operators (LIN, CAN, Ethernet, etc.), I/O drivers, and more by implementing software layers across registers with the microcontroller 1 (Dafang, W., Jiuyang, Z., Guifan, Z. 2010).

1.6.1.2 ECU Abstraction Layer

The ECU abstract layer's main goal is to make ECU-particular administrations more flexibly prominent software layers. This level and its operators are part of the microcontroller and are installed on the ECU equipment. They provide access to all of the ECU's peripherals and devices, which aid functionalities such as memory, communication, and I/O.

1.6.2 SERVICE LAYER

The highest layer in the AUTOSAR core system design is the provider layer. The provider layer creates a working system that operates from the product (S/W) layer to a base microcontroller. The OS serves as a link between the microcontroller as well as the application layer, and it can schedule utility tasks. Network services, storage services, networking services, ECU country management, therapeutics services, and other services are all handled by the carrier layer in BSW. The service layer sits on top of the ECU abstraction layer, in the key neutral of the equipment, and is responsible for providing the application with the required BSW execution. The RTE layer separates the core software from the application software (Kienberger, J. et al 2014).

1.6.3 AUTOSAR RTE LAYER

RTE is a middleware layer of the AUTOSAR software program engineering that provides verbal trade contributions for the application software (Kum, D. et al 2008). It sits between the BSW and software layer.

1.6.4 APPLICATION LAYER

The AUTOSAR programming structure's application layer facilitates the implementation of customized functionalities. The application layer is made up of client-defined programming segments that communicate with sensors and actuators using BSW. It also includes a variety of software components and applications that perform precise tasks according to instructions. The AUTOSAR programming layer is made up of three parts: (1) application programming segments, (2) programming segment ports, and (3) port interfaces. AUTOSAR ensures consistent interfaces for utility layer programming program components, and the utility layer programming program components assist in the development of basic applications to support automobile capabilities. Methodologies to use a virtual Function Bus for specific ports promote communication between computing components (Dafang, W. et al 2010). Furthermore, these ports make it easier for computing components and BSW to communicate. The abovementioned structure of AUTOSAR is its traditional phase, which reinforces ongoing assurance requirements and imperatives (Webers, W., Thörn, C., & Sandkuhl, K. 2008). The critical stage is suitable for supporting purposes in the field of system administration and well-being because of the microcontroller, which allows ECUs to handle automobile sensors and devices.

1.7 ADAPTIVE AUTOSAR

The adaptive platform's improvement is becoming increasingly important and urgent. The truly automatic essence of driving is a true aim in this. When vehicles become halfway responsible of driving, it is essential to have an immediate conversation with the system and cloud servers (Singh, G., Kamath, N., & Sharma, R. K. 2018). In addition, related vehicles and V2X operations necessitate collaboration with the off-board systems. As a result, the on-board discussion must be extremely impenetrable and aid mobile integration, non-AUTOSAR structure incorporation, and cross-domain computing platform levels. Cloud-based administrations necessitate exact protection measures as well. This allows for remote notifications, such as those received over the air. AUTOSAR recently standardized the AUTOSAR adaptive platform to aid in the consistent distribution of client applications and to provide the proper environment for those that need high-end computing (Aust, S. 2018). This is a computer that runs under the POSIX standard. Its most important feature is administration-based correspondence. The adaptive platform is made up of specifications and code. In Compared to Traditional AUTOSAR Adpative AUTOSAR designed and implemented which shortens approval times and reveals secret concepts. It is open to all of the organization's associations.

The AUTOSAR consortium has provided a game-evolving arrangement, which provide efficient work with less time and money contributed than previously. This manner gave OEMs the freedom to choose software program from more than one provider and they suggest higher-quality goods. AUTOSAR's equipment and techniques significantly increase the quality of embedded programming. AUTOSAR's concept is simply the start of a long activity that provides time to implement. Patterns can be used by the automotive industry to refine this software application and enhance its benefits.

From 2003 to 2015, Classic AUTOSAR evolved into a mounted platform that performed admirably well while running 60–80 ECUs in an automobile. Electrification surged with the advancement of Internet of Things (IoT)-based vehicle attributes like V2X network and autonomous driving, and that as an outcome, a huge demand for assisting features and devices was generated in the sector (Swetha, S., & Sivakumar, P. 2021). The current AUTOSAR platform has proven inadequate to accommodate these megatrends, necessitating the creation of stronger new structures and more versatile E/E structure. The adaptive AUTOSAR architecture was implemented to support these features, and the first release of the AUTOSAR scalable model took place in 2017. A central software server that supports high-performance computing comes with an adaptive AUTOSAR structure. Ethernet-based ECUs support real-time functionality in this unit. The adaptive AUTOSAR is modular, with a versatile architecture that allows it to alter functions during the vehicle's lifecycle. This enables OEMs to introduce state-of-the-art software elements in a vehicle and substitute it in the environment as required. AUTOSAR versatile engineering is based on all cutting-edge vehicle applications such as infotainment, V2X, prescient operation, vehicle applications, ADAS highlights with RADAR and LIDAR sensors, camera, zap, map updates, and the sky is the limit from there. The adaptive platform of AUTOSAR is a stable, robust, and incremental platform that lets OEMs design feature-packed vehicles to direct future automotive trends (Soltani, M., & Knauss, E. 2015). The adaptive platform makes use of smart ECUs and enables upgrading over the life cycle of the vehicle for high-end vehicle purposes. To define the adaptive AUTOSAR platform, its functionality, layout, and its use cases, read this article. For OEMs and their customers, car frameworks have grown dramatically and inclinations such as autonomous driving, prescient assistance, V2X availability, OTA updates, and vehicle jolt are a high concentration. The use cases posed by these trends are reshaping the technical experience across subsystems of cars. While the simple AUTOSAR platform helped OEMs and their suppliers with standardized software systems by using the frequent architecture of ECU software program architecture, elevated reuse, and unified language methodology, it also faced constraints to support complex, efficient, and bendy E/E architecture needed for next-gen automotive trends (Wu, R. et al 2010). Then again, the AUTOSAR flexible stage is adaptable, vigorous, and steady. It provides bendy tech configuration that can be modified over the life cycle of the vehicle to assist self-sufficient driving, V2X networking, and various high-end apps. The AUTOSAR adaptive platform provides software developers with a powerful programming interface called AUTOSAR runtime for ARA. The following are some of the requirements that will enable the AUTOSAR adaptive platform to be used in future automotive applications (Webers, W., Thörn, C., & Sandkuhl, K. 2008).

High-performance computing

1. Easy configuration and diagnosis
2. Faster development cycles/rapid prototyping
3. Multicore processor support
4. Interoperability with classic AUTOSAR platform

1.7.1 SERVICE-ORIENTED ARCHITECTURE (SOA)

AUTOSAR adaptive platform focuses on flexibility and scalability and is for that reason designed over SOA. SOA approves designing bendy ECUs software that helps in decreasing the complexity and improving reusability and portability of the software program components.

1.7.2 PROGRAMMING LANGUAGE

Adaptive AUTOSAR relies upon on C 14 language standard. Usually, C language is the top focal point to boost any vehicle application; however, the complexity of the adaptive platform led to the adoption of C. Compared to C language, C offers a higher mechanism for records encapsulation and helps massive and allotted systems in a higher manner.

1.7.3 OS

AUTOSAR adaptive platform runs on POSIX-based (PSE51). POSIX stands for "Portable Operating System Interface" through which one can access the OS. Through PSE 51, about 300 APIs can be used, which allows higher portability. POSIX-based totally (PSE51) additionally helps preemptive multitasking and allows a dynamic variety of tasks.

1.8 ADAPTIVE AUTOSAR ARCHITECTURE

Adaptive AUTOSAR architecture consists of four most important layers:

1. **Hardware/Virtual Machine:** This is the first layer of the architecture and is referred to as a digital machine, on which the adaptive AUTOSAR basis is fashioned and ARAs.

2. **Adaptive AUTOSAR Foundation:** The adaptive AUTOSAR basis carries the simple offerings for the applications, which abstracts the hardware and provides an opportunity for applications to run efficiently.
3. **Adaptive AUTOSAR Services:** Adaptive AUTOSAR services manage the conversation between the adaptive AUTOSAR applications.
4. **ARAs:** The adaptive utility runs on top of the architecture and uses the services supplied by using ARA (AUTOASR runtime for ARAs) as shown in Figure 1.2. ARA affords APIs for adaptive AUTOSAR services like software configuration management, security management, and diagnostics.

The features and architecture mentioned above explain the technical background of the AUTOSAR adaptive platform to support future use cases of automotive trends mentioned below.

1.8.1 Autonomous Driving

To plan the points of independent driving, automobiles have to have sensible ECUs that can deal with the giant quantity of data. Autonomous motors have to communicate with the changing surrounding environment, manipulate alerts from the hundreds of in-vehicle sensors, and deal with the massive data that flows in the clever ECUs. Intelligent ECUs require in-vehicle software to be up to date in the course of a vehicle's existence cycle due to evolving exterior structures and increased functionality. Autonomous automobiles of stage four and level five (partially or independent vehicles) use tremendously complex algorithms, maps, and sensor fusion to feature an adaptive platform that helps in enabling them all (Parekh, T. et al 2021).

1.8.2 Connected Vehicles

Vehicle connectivity is the next massive aspect that is rising exceptionally. Vehicle-to-vehicle (V2V), V2X, remote diagnostics, and cloud-based analytics are some of

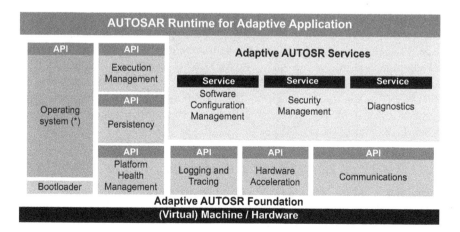

FIGURE 1.2 AUTOSAR runtime for adaptive applications

the hot tendencies in the linked vehicle paradigm (Subburaj, S. D. R. et al 2021). V2X systems require invulnerable communication with different cars and off-board systems, which might also be non-AUTOSAR systems. Next-gen motors will be related to different vehicles, smartphones, traffic infrastructure, etc., and in-vehicle V2X applications will be required to be up to date OTA; this is the place the AUTOSAR adaptive platform assists.

1.8.3 VEHICLE ELECTRIFICATION

The other rising vehicle tendencies like automobile electrification require a complete system-level approach in design, as its single function can create an effect on other necessary functions. The features such as EV charging, charger conversation, and hybrid electric powered motors are related to functionalities of electric cars only and wanted to be addressed through sensible ECUs distinctively than in gasoline-powered vehicles. The electric powered automobiles require extraordinary units of ECU software and options when it comes to V2V, V2X connectivity, physique area controller, and OTA software program updates. The AUTOSAR is the solely successful platform that can meet the incremental requirements of electric automobiles in communication, connectivity, and other areas.

1.8.4 CLASSIC AND ADAPTIVE AUTOSAR ARE COMPLEMENTARY

While AUTOSAR traditional platform supports basic car features such as ignition, the front, and rear mild management, engine control, torque control, etc., AUTOSAR adaptive platform is an evolving platform, which requires excessive overall performance computers that are embedded with clever ECUs to empower infotainment, HMI, ADAS functions, connectivity, and more (Sivakumar, P., Devi, R. S., Lakshmi, A. 2020). The performance of updating the ECU software program over the air will remove the probability of introducing new ECUs to the vehicle. The AUTOSAR adaptive platform has given the advantage to OEMs to launch their high-end vehicles with minimal functionalities. The adaptive platform's scalability and versatility allow automobiles to enhance their purposes and aspects over the vehicle's life cycle, which is how they improve their efficiency eInfochips of AUTOSAR facilities. As an AUTOSAR partner, eInfochips (An Arrow Company) assists Tier 1 suppliers and automobile OEMs with a variety of AUTOSAR-related services, including the creation, implementation, and testing of AUTOSAR simple software modules, the AUTOSAR stack software's implementation from the suppliers, platform software upgrades to latest AUTOSAR specifications as per requirements, and much more (Jayan, J., & Srinivasan, G. 2019).

1.8.5 IMPORTANCE OF AUTOSAR IN CAR INDUSTRY

AUTOSAR enables new digital systems to be introduced that can enhance protection, environmental friendliness, and overall efficiency. The common strategy is to prepare the sector for future applied sciences and reduce fees without losing results.

1.8.6 AUTOSAR-INCORPORATED APPLICATIONS

AUTOSAR is common in the industry for the communication of automotive systems. AUTOSAR specifications describe a layer of BSW that includes services that communicate with precise hardware but have a general application interface. EMCOS AUTOSAR is the eMCOS-compliant AUTOSAR Classic Platform profile, a real-time running computer (real-time systems or embedded systems with integrated RTOs) that was once the first such product available on the market to provide flexible support ranging from single-core to multi-core processors. A standardized interface is an interface that is predefined in the C language by using the AUTOSAR specification as an API. It is used in an ECU between BSW modules, between the RTE and the operating machine, or between the RTE and the BSW module. AUTOSAR is distinct from all other embedded programming paradigms. The reason for the distinction is that in uniform layers, distinguishing the output from complete ECU is needed. Besides, via the modules or software program components, this split operation often determines common interfaces. This standardization has a number of advantages, including scalability to individual devices, acknowledgment of practicable safety requirements, integration of multiple providers' functional fashions, and reusable software components. Aside from that, AUTOSAR is public architecture. A certificate from the board is required to be used for business purposes. A developer cannot sell software under the AUTOSAR brand without the certificate from the board.

1.8.7 ORGANIZATION

- A key associate
- Strategic associate
- Exclusive partnership
- Associate partner
- Development partner

The founding partners BMW, Bosch, Continental, Daimler AG, Ford, General Motors, PSA Peugeot Citroen, Toyota, and Volkswagen are among the attendees' key associates. These companies are in charge of the AUTOSAR improvement partnership's organization, administration, and management. The Executive Board establishes the standard method and roadmap within this core. The steering committee oversees non-technical operations such as partner admission, public family members, and contractual issues. For that reason, the Chairman and Deputy Chairman, each designated for a year, represent the Steering Committee. The contact with the outdoor environment is taken over by the AUTOSAR spokesperson. Strategic associates are selected from the circle of premium partners for a two-year term and assist the undertaking chief group in a variety of technological, operational, and day-to-day processes. They also provide new strategic inputs to the project manager round. Participants in the standard and community development contribute to work applications that are organized and supervised by the Project Manager Team, which was formed with the help of the core partners. The popular files AUTOSAR has already released are being used by associate partners. Attendees are actively working on academic programs and non-profit projects. The AUTOSAR production alliance now includes over 270 organizations as of mid-2019.

1.9   ADVANTAGES

- Hardware and software are largely unrelated to one another.
- Horizontal layers will decouple development (through abstraction), reducing development time and costs.
- Software reusability improves consistency and performance.
- Create a development distribution system among suppliers.
- Improved design versatility allows you to compete on creative features.
- Make program and device integration easier.
- Lower the total cost of software creation.
- Enable more efficient variant handling.
- Share software modules between OEMs.
- Improve application development performance.
- Come up with new business models.
- Work in the planning process.
- Integrate software into a larger tool environment.
- Use normalized applications to enable new business opportunities.
- Gain a clear understanding of how automotive software is created.

1.10   APPLICATIONS

- In 2008, the first automobiles with AUTOSAR technology were introduced to the market. Today, the brand is well-known, and AUTOSAR products are successfully used in a variety of vehicle ventures. The large proportion of the world's automakers is AUTOSAR partners, and the significant proportion of them are either using or planning to use AUTOSAR technology. Auto manufacturing by AUTOSAR OEM associates accounts for roughly 80% of global demand.
- Because of the standardization, there would be more component reuse. Expenses for skill enhancement can be spread out over a longer period of time. Another issue that reduces costs is the interchangeability of supplier solutions. As a result, there is more opposition.
- AUTOSAR emphasizes on technological advancements rather than business models. In the marketplace, there will be some answers to this query.

1.11   CONCLUSION

AUTOSAR encourages a broad impact on the advancement of electronics automation. It assembles electronics, OEMs, semiconductors, and a variety of automation companies to produce a variety of creative products. Elucidation suggested by AUTOSAR methodology extensively reconciles with modern evolution since enhanced effectiveness with considerably diminished time and money plunged provides the liberty to OEMs to select software alternatives from multiple suppliers.

REFERENCES

(Devi, R. S., Sivakumar, P., & Balaji, R. 2018) Devi, R. S., Sivakumar, P., & Balaji, R. 2018. AUTOSAR Architecture Based Kernel Development for Automotive Application. In International Conference on Intelligent Data Communication Technologies and Internet of Things (pp. 911–919).

(Arts, T., Hughes, J., Norell, U. 2015) Arts, T., Hughes, J., Norell, U., & Svensson, H. 2015. Testing AUTOSAR Software with QuickCheck. In 2015 IEEE Eighth International Conference on Software Testing, Verification and Validation Workshops (ICSTW) (pp. 1–4).

(Kienberger, J., Minnerup, P., Kuntz, S 2014) Kienberger, J., Minnerup, P., Kuntz, S., & Bauer, B. 2014. Analysis and Validation of AUTOSAR Models. In 2014 2nd International Conference on Model-Driven Engineering and Software Development (MODELSWARD) (pp. 274–281).

(Parekh, T., Kumar, B. V., Maheswar, R. 2021) Parekh, T., Kumar, B. V., Maheswar, R., Sivakumar, P., Surendiran, B., & Aileni, R. M. (2021). Intelligent Transportation System in Smart City: A SWOT Analysis. In Challenges and Solutions for Sustainable Smart City Development (pp. 17–47). Springer, Cham.

(Mahmud, N., Rodriguez-Navas, G., Faragardi, H. 2018) Mahmud, N., Rodriguez-Navas, G., Faragardi, H., Mubeen, S., & Seceleanu, C. 2018. Power-aware Allocation of Fault-tolerant Multirate AUTOSAR Applications. In 2018 25th Asia-Pacific Software Engineering Conference (APSEC) (pp. 199–208).

(Webers, W., Thörn, C., & Sandkuhl, K. 2008) Webers, W., Thörn, C., & Sandkuhl, K. 2008. Connecting Feature Models and AUTOSAR: An Approach Supporting Requirements Engineering in Automotive Industries. In International Working Conference on Requirements Engineering: Foundation for Software Quality (pp. 95–108).

(Fürst, S., & Bechter, M. 2016) Fürst, S., & Bechter, M. 2016. AUTOSAR for Connected and Autonomous Vehicles: The AUTOSAR Adaptive Platform. In 2016 46th annual IEEE/IFIP International Conference on Dependable Systems and Networks Workshop (DSN-W) (pp. 215–217).

(Subburaj, S. D. R., Kumar, V. V., Sivakumar, P. 2021) Subburaj, S. D. R., Kumar, V. V., Sivakumar, P., Kumar, B. V., Surendiran, B., & Lakshmi, A. N. 2021. Fog and Edge Computing for Automotive Applications. In Challenges and Solutions for Sustainable Smart City Development (pp. 1–15). Springer, Cham.

(Honekamp, U. 2009.) Honekamp, U. 2009. The Autosar XML Schema and Its Relevance for AUTOSAR Tools. IEEE Software, 26(4), 73–76.

(Elbahnihy, A., Safar, M., & El-Kharashi, M. W. 2020) Elbahnihy, A., Safar, M., & El-Kharashi, M. W. 2020. Hardware-accelerated SOME/IP-based Serialization for AUTOSAR Platforms. In 2020 15th Design & Technology of Integrated Systems in Nanoscale Era (DTIS) (pp. 1–2).

(Dafang, W., Jiuyang, Z., Guifan, Z. 2010) Dafang, W., Jiuyang, Z., Guifan, Z., Bo, H., & Shiqiang, L. 2010. Survey of the AUTOSAR Complex Drivers in the Field of Automotive Electronics. In 2010 International Conference on Intelligent Computation Technology and Automation (Vol. 3, pp. 662–664).

(Ryu, H., Jnag, S. Y., & Lee, W. J. 2013) Ryu, H., Jnag, S. Y., & Lee, W. J. 2013. AUTOSAR Unit Testing Approach Based on Virtual Prototyping for Software Components. In 2013 Fifth International Conference on Ubiquitous and Future Networks (ICUFN) (pp. 114–116).

(Sivakumar, P., Vinod, B., Devi, R. 2016a) Sivakumar, P., Vinod, B., Devi, R. S., & Divya, R. (2016). Deployment of Effective Testing Methodology in Automotive Software Development. Circuits and Systems, 7(9), 2568–2577.

(Wu, R., Li, H., Yao, M. 2010) Wu, R., Li, H., Yao, M., Wang, J., & Yang, Y. 2010. A Hierarchical Modeling Method for AUTOSAR Software Components. In 2010 2nd International Conference on Computer Engineering and Technology (Vol. 4, pp. V4–184).

(Singh, G., Kamath, N., & Sharma, R. K. 2018) Singh, G., Kamath, N., & Sharma, R. K. 2018. Implementing Adaptive AUTOSAR Diagnostic Manager with Classic Diagnostics as APIs. In 2018 Second International Conference on Intelligent Computing and Control Systems (ICICCS) (pp. 894–898).

(Aust, S. 2018) Aust, S. 2018. Paving the Way for Connected Cars with Adaptive AUTOSAR and AGL. In 2018 IEEE 43rd Conference on Local Computer Networks Workshops (LCN Workshops) (pp. 53–58).

(Sivakumar, P., Vinod, B., Devi, R. S. 2016b) Sivakumar, P., Vinod, B., Devi, R. S., & Divya, R. 2016. Novelty Testing Measures and Defect Management in Automotive Software Development. *Australian Journal of Basic and Applied Sciences*, 10(1), 607–613.

(Kim, J. W., Lee, K. J., & Ahn, H. S. 2015) Kim, J. W., Lee, K. J., & Ahn, H. S. 2015. Development of Software Component Architecture for Motor-driven Power Steering Control System Using AUTOSAR Methodology. In 2015 15th International Conference on Control, Automation and Systems (ICCAS) (pp. 1995–1998).

(Kum, D., Park, G. M., Lee, S. 2008) Kum, D., Park, G. M., Lee, S., & Jung, W. 2008. AUTOSAR Migration from Existing Automotive Software. In 2008 International Conference on Control, Automation and Systems (pp. 558–562).

(Swetha, S., & Sivakumar, P. 2021) Swetha, S., & Sivakumar, P. 2021. SSLA Based Traffic Sign and Lane Detection for Autonomous cars. In 2021 International Conference on Artificial Intelligence and Smart Systems (ICAIS) (pp. 766–771).

(Jayan, J., & Srinivasan, G. 2019) Jayan, J., & Srinivasan, G. 2019. AUTOSAR Based Dual Core Partitioning for Power Train Application of BMS. In 2019 IEEE International Conference on Intelligent Techniques in Control, Optimization and Signal Processing (INCOS) (pp. 1–6).

(Sivakumar, P., Devi, R. S., Lakshmi, A 2020) Sivakumar, P., Devi, R. S., Lakshmi, A. N., VinothKumar, B., & Vinod, B. 2020. Automotive Grade Linux Software Architecture for Automotive Infotainment System. In 2020 International Conference on Inventive Computation Technologies (ICICT) (pp. 391–395).

(Soltani, M., & Knauss, E. 2015) Soltani, M., & Knauss, E. 2015. Challenges of Requirements Engineering in AUTOSAR Ecosystems. In 2015 IEEE 23rd International Requirements Engineering Conference (RE) (pp. 294–295).

(Dafang, W., Shiqiang, L., Bo, H. 2010) Dafang, W., Shiqiang, L., Bo, H., Guifan, Z., & Jiuyang, Z. 2010. Communication Mechanisms on the Virtual Functional Bus of AUTOSAR. In 2010 International Conference on Intelligent Computation Technology and Automation (Vol. 1, pp. 982–985).

(Sandhya, D. R., Sivakumar, P., & Balaji, R. 2019) Sandhya Devi R.S., Sivakumar P., Balaji R. 2019. AUTOSAR Architecture Based Kernel Development for Automotive Application. In: Hemanth J., Fernando X., Lafata P., Baig Z. (eds) International Conference on Intelligent Data Communication Technologies and Internet of Things (ICICI) 2018. ICICI 2018. Lecture Notes on Data Engineering and Communications Technologies, vol 26. Springer, Cham.

2 Use of Communication Protocols in Automotive Software Development Process

R. S. Sandhya Devi
Department of EEE, Kumaraguru College of Technology,
Coimbatore, India

P. Sivakumar
Department of EEE, PSG College of Technology,
Coimbatore, India

B. Vinoth Kumar
Department of IT, PSG College of Technology,
Coimbatore, India

A. D. Buvanesswaran
PSG College of Technology, Peelamedu, India

CONTENTS

DOI: 10.1201/9781003269908-2

2.1 COMMUNICATION PROTOCOL

In general, communication protocols are used to transfer information from one entity to other or from one to many which are connection together in a common network. This information is transferred in a well-defined format based on the protocol used. This information is not transferred just like that. We have well-defined formats starting from encryption till error recovery methods. Communication takes place between two nodes with a pre-defined format which two devices should be able to understand. Each message should carry a particular meaning whether it is an information or command, and the same should be implicitly decoded by the receiver. This response by the receiver should be as the one intended by the sender. Communication protocols must be common and should be agreed by the stakeholders involved in the communication.

2.1.1 NEED FOR COMMUNICATION PROTOCOL IN AUTOMOTIVE SOFTWARE DEVELOPMENT

Automotive systems need networking to deal with the tedious architecture which connect various parts together. Almost all the car manufacturers get the components needed for their car from other common companies which agreed to supply the components for many OEMs as a common architecture (Rappl, M. et al 2002). Recent inventions like ABS, airbag deployment module need interaction from various parts of the car, which is possible only by networking. We have shifted from mechanical interactions to X-by-wire technologies. The concept of technology replacement should happen only when the latter is more reliable than the former. Starting from sophistication for the passengers till dealing with the exhaust norms laid by the Government, everything needs communication. Also, we cannot have a common networking architecture as different applications need different types of communication as we are dealing with real-time systems.

An automotive system consists of many electronic control units (ECUs) which work together or sometimes independently. Depending on the OEMs an automotive system consists of minimum one ECU to a maximum 70 ECUs. This adds more complexity to the system, particularly in networking part. In order to overcome this complexity, several OEMs have come together to agree upon a common standard architecture for automotive systems, which is AUTOSAR and communication protocols like Flex ray (Sandhya, D. R., Sivakumar, P., & Balaji, R. 2019). In contrary to normal braking and steering using mechanical parts, the X-by-wire technology has used several actuators and sensors to replace the mechanical system. For the communication between the actuator and sensor components, we go for standard protocol.

If we take an automotive system, the requirement starts from the application we use and criticality of that application. This decides how much importance we can give for designing the network. If we divide the automotive system based on the application and purpose, we can go with chassis, power train, body electronics, and infotainment. These subsystems have different networking needs and this has to be addressed with low latency. Another important thing is that, these networks have to work easily together; there should be compatibility between them. Automotive systems are hard real-time systems, so error is not acceptable. And also, a car manufactured today with the existing technologies should be flexible enough to update with the new inventions, as a car runs for a minimum of ten years on road. Ten years is an ample time for inventions. When an automotive system using minimum networks to solve various requirements is appreciable as it reduces complexity and time. In order to reduce this complexity, we have to go for a common networking topology. These networks should support future plug-ins, less time requirement and support "Network in Networks" (Muller, C., & Valle, M. 2011).

2.2 AUTOMOTIVE SYSTEMS

As explained earlier, an automotive system consists of chassis, power train, body electronics, and infotainment in a broader sense. These have to be interconnected using sensors and actuators to work together. This subsystem is not a one-day invention; it is evolved over time, so needs a network of that kind.

2.2.1 HISTORY OF AUTOMOTIVE INVENTIONS

In the beginning, all the automotive components are connected by hydraulics and mechanical connections. As time goes on, these were replaced by wires and cables for error-free and comfort operations. This transformation is done step by step.

Today almost all automotive subsystems depend on electronics, so here we introduced the concept of bus for hassle-free communication. When bus comes into picture, we can utilize the concept of networking among all the dependent components. Advancement towards this is fieldbus which is used for transferring messages from one point to another. This field bus carries information to from an ECU. As of today, almost all the automotive communication takes place through this fieldbus (Chávez, M. L. 2006). In fieldbus, we have to take care of latency, errors, security, etc. The advancement in electronics made the communication between the ECUs, actuators, and sensors easier.

Initially, the OEMs themselves designed the fieldbus for their own dedicated purpose, but as time goes on this R&D in this communication and maintenance took them more time for taking the product to the market, so they went to the subcontractors who has standardized concepts among OEMs and get it from them. Here, OEMs just need to plug and play. As more OEMs are getting these done by subcontractors, standardization was done. In the beginning of 90s control area network (CAN) was introduced, which is a standardized protocol developed by Bosch. After that, CAN became more popular, but CAN has used only two layers in OSI model, so there were several advancements in CAN like Device Net, CAN Kingdom, and CANopen. These protocols used higher layers of OSI model which gives easy maintenances and usability.

As the need pushed them to move form hydraulics and mechanical components to electronic control, they were in need of high-speed bus system. After several rounds of research and implementation in cars, they came with the system called "X-by-wire" system. Reasons for moving from hydraulics to electronics are many, but the most important one will be cost and maintenance of these systems and also advancement needs completely new implementation. Another reason is recycling of heavy hydraulic components are tedious.

2.3   AUTOMOTIVE COMMUNICATION REQUIREMENTS

Automotive communication requirements depend on the features with which the network is associated with. In general, we can classify them as Security, flexibility, bandwidth, determinism, and fault tolerance.

Security—Automotive system nowadays has recent technologies such as engine control, driver assistance, braking mechanism, etc. These are controlled by software. The cars nowadays have Wi-Fi connectivity for communication with the outside world mainly for infotainment. Data is transferred to and from cloud. There is possibility of hacking the bus system and injecting fault codes. This results in serious issues pertaining to the passengers.

Flexibility—Automotive subsystems are implemented based on their criticality as discussed before. Based on this, a subsystem can be event triggered or time triggered. For example, airbag deployment module is event triggered; airbag has to deploy only when there is an impact. On the other hand, most warning systems are time triggered. So, a communication protocol has to be flexible enough to accommodate this. Also, we have several mechanisms like bus arbitration, etc.

Bandwidth—Bandwidth, in general, refers to amount of data transmitted in a bus in a given amount of time. When we take automotive system in specific, the data need to be accurate and for an ECU it has to transmit and receive data to and from all the nodes connected to the ECU, so here the bandwidth need to more. For example, if we take antilock braking system (ABS), for an ABS to work properly, it should get data from several sensors such as engine speed, vehicle velocity, etc. So, these data will be transmitted simultaneously in a single ECU. So, the bus system should be capable enough to accommodate more data at the same time.

Determinism—Automotive communication system should be deterministic, there should be correct reception of messages. The messages should be sent and received in a predicted amount of time. For periodic systems to work, data has to be transmitted in the defined time interval else the system might go wrong. As mentioned earlier, in automotive system, we are dealing with many hard real-time systems, so determinism is a key factor for these systems to work properly (Sivakumar, P. et al 2015).

Fault tolerance—In automotive system, we have many sensors associated with hardware. Hardware components are susceptible to wear and tear. So, they might go faulty at any time, especially in unpredicted situations. Correspondingly, the sensor data goes wrong, which results in error in communication. These errors in communication should be predicted in advance and response should be provided, else the signal goes corrupted. Also, the signal might get lost due to various physical factors; in this situation also, some decision has to be taken, so the system should be fault tolerant.

2.4   TYPICAL SUBSYSTEM MODULES

In automotive system, networking is dependent on many modules; in a broader sense, we have classified them as eight subsystem modules. They are:

Chassis subsystems come under vehicle active safety. We have two subsystems here, electronic stability program (ESP) and vehicle dynamics control (VDC), this is used to assist the driver in terms of dynamics of the car (Devi, R. S., Sivakumar, P., & Sukanya, M. 2018). Engine ECU does this part. To overcome the possibility of over steering, under steering, skidding, we have these systems. ABS also comes under this, to prevent the locking of wheel while applying the brake at high velocity, ABS need feedback communication as it needs the wheel speed for every cycle.

Airbag deployment module comes under passive safety system. Airbag deployment module is a hard real-time system. When a crash occurs, airbag has to be deployed in minimum time, in order to prevent the passenger from hitting in the front. Here from the time of impact, various step by step processes have to be carried out before deployment. First seatbelt status should be checked, then seatbelt pretensioners should be activated, after that only airbag has to be deployed. This has to happen in order; means the communication to and from the airbag deployment module has to be in such a way so that airbag will be deployed successfully.

Powertrain systems include the combustion of the fuel to produce power, till the power reaches the wheel. This path needs communication at a greater co-ordination. There should not be any loss as such. From the flywheel to the differential thorough the gear box, there should be hassle-free transfer of power. This involves co-ordination in injector timing, cam timings, exhaust gas recirculation, etc.

Body electronics does not come under safety, critical system. Seat belt adjustment, wiper control, headlight control, window control are some of the human-machine interface (HMI) which are soft real-time systems (Sivakumar, P. et al 2016). So, networking in these areas can be cost-effective, unless importance is given to it for sophistication. However, automation in body electronics requires some dedicated network.

X-by-wire technology, as already explained, is the concept of replacing hydraulics and mechanical components with computational electronics. So here, networking is necessary. We have steer-by-wire, shift-by-wire, brake-by-wire, throttle-by-wire. These are the subsystems which need dedicated communication to work efficiently.

Infotainment systems include music player, speakers, voice control, games etc. These are not safety critical systems, so communication in these systems need not be worried about in terms of fault tolerance or latency.

Telematics includes connecting the vehicle to the external world. This includes GPS, traffic management, vehicle communication, etc. These systems need secure communication as they are prone to attack from the outsiders. There are many technologies for secure communication with the Internet.

Advanced driver assistance system (ADAS) assists the driver in driving and parking functions. This is the base for autonomous driving field (Xu, Y.N. et al 2008) (Swetha, S., & Sivakumar, P. 2021). This is done with the help of sensors and cameras positioned at the various points in the vehicle. Some if its features include lane departure warning, break assistance, emergency braking, etc. (Broy, J., & Muller-Glaser, K. D. 2007). To be specific, ADAS needs proper communication only, from the sensor to the system, from the system to the driver.

All the above-mentioned systems come under driving systems, which is needed in a car to work. Another important system is on board diagnostics (OBD). OBD is a self-diagnostic technology, where a car records the error and stores it as a specific code related to the error. This we call it as diagnostic fault codes (DFC). These DFCs are retrieved by the technician when the vehicle is taken for service. Reaching each subsystem and debugging is a time consuming and tedious job for the technicians. OBD also needs communication network from almost all the subsystems in the car, to record the fault codes accurately.

2.5 AUTOMOTIVE COMMUNICATION TECHNOLOGIES

We have various automotive technologies used for the communication needs which are mentioned above. Starting from the ancient technologies till the current trend, some of them will be discussed in this section

2.5.1 WIRED TECHNOLOGIES

Automotive manufacturers are using various technologies for communication. Some of the most commonly used ones are CAN, LIN, MOST, Flexray. We will look into these in detail and will see a brief of other technologies.

2.5.2 CONTROL AREA NETWORK (CAN)

Robert Bosch developed CAN in 1985 for in-vehicle communication (Robert Bosch Gmbh 2013). The development is a result of series of research in areas of automotive, since the networks in car starts increasing due to increase in technologies. At first, there are direct wire connections between components so complexity was less. But as the functionalities increase, direct wire connection is not possible as the connection becomes bulkier and complex for operations and also debugging typical network without CAN and With CAN as shown in Figure 2.1. In order to overcome this, Bosch developed CAN.

CAN Benefits: CAN offers a cost-effective, durable network which enables many CAN devices to connect with each other. The use of this is that every device in the system can have single CAN interface better than analogue and digital inputs to ECUs. In cars, this reduces overall cost and weight. Each message has a priority, in case if two attempt to send messages at the same time, the one with the higher priority is sent and the one with the lower priority is delayed.

Each one of the devices have CAN controller chip on the network and is therefore intelligent. All the computer devices see all the set messages.

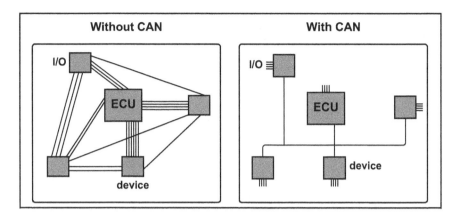

FIGURE 2.1   Control area network. (National Instrument white paper. 2020)

To carry out error testing on each frame's contents, the CAN specification contains a cyclic redundancy code (CRC). Both node frames with errors and an error frame may be transmitted to indicate the error in the network. The controller differentiates between global and local errors, and incase more errors arise, corresponding nodes stop sending errors or they get themselves disconnected (Wittmann, R., & Zitterbart, M. 2000).

CAN Physical Layers: CAN has multiple distinct physical layers that can be used. Some elements of the CAN network are defined by these physical layers, such as transmission rate, signaling models, cable properties, baud rate, etc.

The physical layers that are widely used are discussed below:

Low-Speed/Fault-Tolerant CAN Hardware: Normally, CAN networks are often designed with two wires, can communicate with devices at speeds up to 125 kbit/s, and provide fault-tolerant transceivers. Other names include CAN B and ISO 11898-3 for low-speed/fault-tolerant CAN. Comfort devices include traditional low-speed/fault-tolerant devices in a car.

High-Speed/FD CAN: The popular physical layer by far is the high-speed CAN. With two wires, high-speed CAN networks are implemented and enable communication at transfer rates of up to 1 Mbit/s Other high-speed CAN names include CAN C and 118982 ISO.

Software-Selectable CAN Hardware: With National Instruments CAN hardware products, we can build CAN network either of the onboard transceivers (high-speed, low-speed/fault-tolerant, or single-wire CAN) to use the software-selectable CAN modules. Multiple transceiver modules will be a better solution for applications that use a wide range of communication protocols (Johansson, K. H., Törngren, M., & Nielsen, L. 2005).

Single-Wire CAN Hardware: Communication speed can be up to 33.3 kbit/s (88.3 kbit/s in high-speed mode) in single wire CAN communication networks. Other names include SAE-J2411, CAN A, and GMLAN for single-wire CAN.

Working of CAN Communication: CAN networks do not have any master for controlling the communication. Instead, each node can write CAN frame into the network, after checking whether the bus is busy. While writing into the network, no address of the receiver is included in the frame. Alternatively, an arbitration ID is included, which will be read by all the CAN nodes. These nodes will decide whether to accept the frame, based on the arbitration ID (Sivakumar, P. et al 2015) (Paul, A. et al 2016).

2.5.3 LOCAL INTERCONNECT NETWORK (LIN)

LIN Topology and Behavior: LIN bus is constituted by a single master device and multiple slave devices. Node capability file defines each node's behavior. A device generator parses the LDF to automatically generate the specified behavior within the desired nodes. The master node master task begins transmitting headers to the bus at this stage, and all the cluster slave tasks (including the master node's own slave task) react, as defined in the LDF.

LIN Error Detection and Confinement: The LIN 2.0 specification notes that the slave tasks can manage error detection and that there is no need for error monitoring by the master task. The LIN 2.0 specification does not require the handling within one LIN frame of multiple errors or the use of error counters. Upon encountering the first error in a frame, the slave task aborts frame processing before the next break-sync sequence (transmitted by the master in the next header) is observed. If the log bus error attribute is set to true, the read queue records a bus error frame.

2.5.4 MOST

For distinct functions, a MOST network must have a range of masters. Masters may be housed in the same unit.

Timing Master: Controls network timing and, therefore, synchronization between devices. Network master sets up the network and assigns system addresses. Connection master sets up synchronous channels of communication between devices.

Power master
Control power supplies
Power-up and
shut down handles

A MOST system consists of three parts. Network Services Physical Interface—These services are managed by a network interface controller (NIC). Modern NICs have a processor built in and are called intelligent NICs, INICs.

2.5.4.1 Function Blocks (F Blocks)

These take care of the services that Function Blocks (FBlocks) can provide to the system. A MOST system is not connected to a bus in the common sense. It has an in port and out port and passes the information from the in port to the out port. FBlocks can have functions with two different targets. The MOST t(tk) FB locks application can be of three types. Controllers and slaves control one or more F Blocks. The MOST framework –HMIs-Human Computer

Interface-Used for user-device interaction. Through the F blocks, communication between devices takes place. Without the sender understanding the address of the receiving F block, the communication can take place. In the sending device, this is achieved by so-called shadows. If the device includes more than one receiving FBlock of the same kind, the connection will be much more complicated and the device will have to handle the transfer and addressing of the NIC.

2.5.5 FLEXRAY

BMW and Daimler-Chrysler after examination found that the existing technologies will not support the automobile industry for future development in technologies especially X-by-wire systems (Mane, S. P., Sonavane, S. S., & Shingare, P. P. 2011). So, they found that Flexray offers a solution for implementing X-by-wire systems by removing some of the fieldbuses currently in use, thus, decreasing the overall network density. Basically, today all car manufacturers joined this consortium, and in mid-2004, the protocol specification was made public. In future automotive systems, FlexRay is used in ECU communication which need faster data exchange.

2.5.6 BYTEFLIGHT

Byteflight was introduced in 1996 by BMW, and then further developed by BMW, ELMOS, Infineon, Motorola, and Tyco EC. Safety-critical systems need more bandwidth, which may now be satisfied by CAN but not in the future. So, Byteflight provides a solution for this. ABS is one of the applications. When Byteflight was developed, the key requirements were versatility, increased bandwidth than CAN. Byteflight supports network speeds of up to 10 MBps. For X-by-wire systems, Byteflight is a nominee. It has, however, been expanded to be part of the protocol of FlexRay.

2.5.7 OTHER TECHNOLOGIES

CAN, LIN, and Byteflight are most commonly applied in data exchange used for chassis, power train, air bag, and electronics for physics and remedies, and diagnostics. These are the first auto-rationale networking implementations in history, but many innovations have been used over the years. Some were recurrent, while others were eventually removed.

The Distributed Systems Interface, a master/slave network with verbal exchange rates of up to 150 Kbps, used for security-related applications.

Safe-by-Wire is a master/slave network used for airbag control. Safe-by-wire has elements taken from CAN and 150 Kbps conversation rates are supported. Since CAN and LIN are no longer considered secure enough for airbag control, this verbal exchange protocol was designed and developed by the Sound-by-Wire Consortium.

Motorola Interconnect (MI) In the experience that it is a simple low-fee master/slave network specially designed for smart sensors, MI is comparable to LIN. However, in today's car systems, LIN is closed to being the world general.

Initially, the combination of computing devices with multimedia gadgets was the position of multimedia and infotainment.

Mobile Multimedia Connection (MML Bus) by Delphi Automotive Systems'. It is providing 100 Mbps communication and plug-and-play feature of the master/slave optic network.

IDB-1394 (Automotive Firewire) was initially used to connect PC computers, but also to try to enter the automotive industry.

Domestic Digital Bus (D2B) via the Optical Chip Consortium. It is a group of ring/star optics providing up to 20 Mbps of communication. In some Mercedes-Benz models, D2B is included.

USB as Firewire, originally used in the PC market now trying to reach to the automotive market.

2.5.8 An Example Automotive System

Communications are split into three categories: (1) powertrain and chassis, (2) body electronics, and (3) infotainment. The network infrastructure of Volvo XC90 as shown in Figure 2.2, it consists of nearly 40 control units, and the CAN occupies the major part which is used for interconnecting these ECUs. The Different ECUs used in each blocks as shown in Table 2.1.

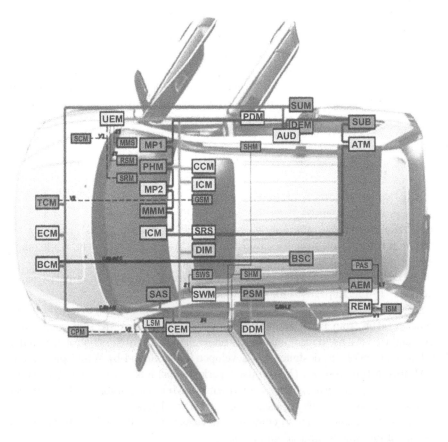

FIGURE 2.2   Network infrastructure of Volvo XC90. (Nolte, T., Hansson, H., & Bello, L. L. 2005)

TABLE 2.1
The Different ECU used in Volvo XC90

Block	Powertrain and Chassis	Block	Infotainment
TCM	Transmission control module	ICM	Infotainment control module
ECM	Engine control module	**Block**	**Body electronics**
BCM	Brake control module	DDM	Driver door module
BSC	Body sensor cluster	REM	Rear electronic module
SAS	Steering angle sensor	PDM	Passenger door module
SUM	Suspension module	CCM	Climate control module
DEM	Differential electronic Module	ICM	Infotainment control module
Block	**Infotainment**	UEM	Upper electronic module
AUD	Audio module	DIM	Driver information module
MP1	Media player 1	AEM	Auxiliary electronic module
MP2	Media player 2	SRS P	Supplementary restraint system
PHM	Phone module	PSM	Passenger seat module
MMM	Multimedia module	SWM	Steering wheel module
SUB	Subwoofer	CEM	Central electronic module
ATM	Antenna tuner module		

Source: Nolte, T., Hansson, H., & Bello, L. L. (2005).

There are many cars which uses 100 ECUs in total. It is becoming more and more difficult to incorporate these subsystems into the communication networks. Volvo uses the Volcano definition in the case of the XC90. It is also possible to perform a timings analysis of the method by using the Volcano instruments.

2.6   WIRELESS TECHNOLOGIES

For intra-vehicle and inter-vehicle communication, there are many communication protocols that can be used for various applications. Extra- and extra-transportable devices, such as cell phones, transportable GSM devices and laptop computer systems, may want to make the most of the vehicle communication when looking at in-vehicle communications. Also, some new roles will leverage the possibility of inter-vehicle communications, e.g., vehicle-to -vehicle and vehicle-to-roadside communications.

2.6.1   ZIGBEE

The new low-cost and low-power wireless PAN standard is ZigBee (IEEE 802.15.4) designed for applications like sensor and controls. Given the number of proprietary systems with low data rates designed to meet the above specifications, there were no criteria that fulfilled them. In addition, the use of such devices has posed major interoperability concerns that ZigBee technology addresses, offering solution for sensor and control applications. In December 2004, the first ZigBees specification for wireless data communications was ratified by the ZigBee Alliance (with over

120 company members). ZigBee offers network speeds of up to 250 Kbps and is intended to be widely used for control and monitoring purposes in the automotive industry as a sensor network.

2.6.2 BLUETOOTH

Bluetooth (IEEE 802.15.1) officially offers up to 3 Mbps (Bluetoothv2.0) network speeds. Bluetooth technology offers low-power, low-cost, short-range communication.

As a potential automotive wireless networking technology, the automotive environment is also very enticing.

The Bluetooth Special Interest Group (SIG) founded the Car Working Group in December 1999 in response to this interest. The hands-free profile was the first of many requirements planned from the Car Working Group for the application stage (Kovačević, J., Kaprocki, N., & Popović, A. 2019).

The Bluetooth SIG set out a three-year plan for potential bluetooth enhancements in November 2004. Quality of Service, security, power consumption, multicast capabilities, and privacy improvements are prioritized targets. It is expected that long-range performance improvements will increase the range of bluetooth-enabled sensors.

2.6.3 OTHER TECHNOLOGIES

Wi-Fi stands for wireless fidelity and for every form of IEEE 802.11 network is the general term. Wi-Fi is used, for example, by the Car2Car Consortium, a non-profit organization initiated by European vehicle manufacturers, for inter-vehicle communications (Devi, R. S., Sivakumar, P., & Balaji, R. 2018) (Sivakumar, P. et al 2020). Advanced drive assistance to minimize the number of incidents, decentralized floating car data to enhance local traffic flow and performance, and comfort and business applications for user communications and information services to drive and passengers are applications here. For example, the European network-on-wheels (NoW) project is a research project working in this field.

UWB (IEEE802.15.3a), or ultrawide band, has become rival to the IEEE 802.11 standards. UWB supports high bandwidth communication. Collision detection systems and suspension systems reacting to road conditions are other possible vehicle applications that could be assisted by UWB. However, as UWB is a young technology, these applications are not yet accessible.

2.7 CONCLUSION

We have discussed about the trends in technologies of automotive communication, the recognition of typical requirements and novel vehicle technology specifications and addressing to what extent networking is available and upcoming technologies are capable of satisfying such demands. The next steps in automotive communications were discussed, focusing on X-by-wire systems and applications for wireless communications. Today, one of the greater problems is interconnecting potentially modern automotive architecture networks that are heterogeneous. This will be answered by implementing standardized middleware technologies.

REFERENCES

(Broy, J., & Muller-Glaser, K. D. 2007) Broy, J., & Muller-Glaser, K. D. 2007. The Impact of Time-Triggered Communication in Automotive Embedded Systems. In 2007 International Symposium on Industrial Embedded Systems (pp. 353–356).

(Chávez, M. L. 2006) Chávez, M. L. 2006. Fieldbus Systems and Their Applications 2005. In Proceedings Volume from the 6th IFAC International Conference.

(Devi, R. S., Sivakumar, P., & Balaji, R. 2018) Devi, R. S., Sivakumar, P., & Balaji, R. 2018. AUTOSAR Architecture Based Kernel Development for Automotive Application. In International Conference on Intelligent Data Communication Technologies and Internet of Things (pp. 911–919).

(Devi, R. S., Sivakumar, P., & Sukanya, M. 2018) Devi, R. S., Sivakumar, P., & Sukanya, M. (2018). Offline analysis of sensor can protocol logs without can/vector tool usage. International Journal of Innovative Technology and Exploring Engineering, 8(2S2), pp. 225–229

(Johansson, K. H., Törngren, M., & Nielsen, L. 2005) Johansson, K. H., Törngren, M., & Nielsen, L. 2005. Vehicle applications of controller area network. In Handbook of Networked and Embedded Control Systems (pp. 741–765). Birkhäuser, Boston.

(Kovačević, J., Kaprocki, N., & Popović, A. 2019) Kovačević, J., Kaprocki, N., & Popović, A. 2019. Review of Automotive Audio Technologies: Immersive Audio Case Study. In 2019 Zooming Innovation in Consumer Technologies Conference (ZINC) (pp. 98–99).

(Mane, S. P., Sonavane, S. S., & Shingare, P. P. 2011) Mane, S. P., Sonavane, S. S., & Shingare, P. P. 2011. A Review on Steer-by-Wire System using Flexray. In 2011 2nd International Conference on Wireless Communication, Vehicular Technology, Information Theory and Aerospace & Electronic Systems Technology (Wireless VITAE) (pp. 1–4).

(Muller, C., & Valle, M. 2011) Muller, C., & Valle, M. 2011. Design and Simulation of Automotive Communication Networks: the Challenges. e & i Elektrotechnik und Informationstechnik, 128(6), 228–233.

(National Instrument White paper. 2020) Controller Area Network (CAN) Overview. National Instrument White paper, 06, 2020. Available (https://www.ni.com/en-in/innovations/white-papers/06/controller-area-network--can--overview.html)

(Nolte, T., Hansson, H., & Bello, L. L. 2005) Nolte, T., Hansson, H., & Bello, L. L. 2005. Automotive Communications-past, Current and Future. In 2005 IEEE Conference on Emerging Technologies and Factory Automation, Vol. 1, pp. 8–992.

(Paul, A., Chilamkurti, N., Daniel, A. 2016) Paul, A., Chilamkurti, N., Daniel, A., & Rho, S. 2016. Intelligent Vehicular Networks and Communications: Fundamentals, Architectures and Solutions. Elsevier.

(Rappl, M., Braun, P., von der Beeck, M. 2002) Rappl, M., Braun, P., von der Beeck, M., & Schröder, C. 2002. Automotive software development: A model-based approach (No. 2002-01-0875). SAE Technical Paper.

(Robert Bosch Gmbh 2013) Robert Bosch Gmbh. 2013. Bosch Automotive Electrics and Automotive Electronics. In Systems and Components, Networking and Hybrid Drive, 5th edition. Springer Vieweg.

(Sandhya, D. R., Sivakumar, P., & Balaji, R. 2019) Sandhya Devi R.S., Sivakumar P., Balaji R. 2019. AUTOSAR Architecture Based Kernel Development for Automotive Application. In Hemanth J., Fernando X., Lafata P., Baig Z. (eds) International Conference on Intelligent Data Communication Technologies and Internet of Things (ICICI) 2018. Lecture Notes on Data Engineering and Communications Technologies, vol 26. Springer, Cham.

(Sivakumar, P., Devi, R. S., Lakshmi, A 2020) Sivakumar, P., Devi, R. S., Lakshmi, A. N., VinothKumar, B., & Vinod, B. 2020. Automotive Grade Linux Software Architecture for Automotive Infotainment System. In 2020 International Conference on Inventive Computation Technologies (ICICT) (pp. 391–395).

(Sivakumar, P., Vinod, B., Devi, R. S. 2015) Sivakumar, P., Vinod, B., Devi, R. S., & Rajkumar, E. J. 2015. Real-time task scheduling for distributed embedded system using MATLAB toolboxes. Indian Journal of Science and Technology, 8(15), 1–7.

(Sivakumar, P., Vinod, B., Devi, R. S. 2016) Sivakumar, P., Vinod, B., Devi, R. S., & Divya, R. S. (2016). Deployment of effective testing methodology in automotive software development. Circuits and Systems, 7(9), 2568–2577.

(Swetha, S., & Sivakumar, P. 2021) Swetha, S., & Sivakumar, P. 2021. SSLA Based Traffic Sign and Lane Detection for Autonomous cars. In 2021 International Conference on Artificial Intelligence and Smart Systems (ICAIS) (pp. 766–771).

(Wittmann, R., & Zitterbart, M. 2000) Wittmann, R., & Zitterbart, M. 2000. Multicast Communication: Protocols, Programming, & Applications. Elsevier.

(Xu, Y. N., Jang, I. G., Kim, Y. E 2008) Xu, Y. N., Jang, I. G., Kim, Y. E., Chung, J. G., & Lee, S. C. 2008. Implementation of FlexRay Protocol with an Automotive Application. In 2008 International SoC Design Conference (Vol. 2, pp. II-2–II-25).

3 Bootloader Design for Advanced Driver Assistance System

R. S. Sandhya Devi
Department of EEE, Kumaraguru College of Technology,
Coimbatore, India

B. Vinoth kumar
Department of IT, PSG College of Technology,
Coimbatore, India

P. Sivakumar
Department of EEE, PSG College of Technology,
Coimbatore, India

A. Neeraja Lakshmi
Department of EEE, PSG College of Technology,
Coimbatore, India

R. Tripathy
Product Engineering Services, KPIT Technologies Ltd.,
Pune, India

CONTENTS

DOI: 10.1201/9781003269908-3

31

3.1 INTRODUCTION

Each electronic control unit (ECU) in an automobile is responsible for the execution of a particular program. For instance, an ECU antilock braking system makes sure brakes are not stuck during braking. The ABS software program in the ECU hardware is capable of recording the vehicle speed as an input and is designed based on this information to minimize the brake on the wheels.

Bootloader is the software algorithm which is performed during device boot. Host of functionalities are provided by automotive ECUs (control units). Such technologies and functionalities have become highly sophisticated and complex. To the automotive original equipment manufacturers (OEMs) and suppliers, it has become crucial to ensure that these software-driven control units are still working in a secure and productive environment. This will only be done if the ECUs have the current and patched version of the software and security updates inside the car. Hence, the framework developed and ported on the MCU platform will also be modified very regularly, either from a remote location or at the service station. This task of initiating the ECU software upgrade was assigned to the bootloader program, which occupies the ECU ROM.

When looking at a luxury vehicle, though, one can note that it has almost 100 million lines of software code. But this is very huge compared to an aircraft, which has only six million lines of code. Therefore, owing to these large amounts of electronic code, the firmware upgrade of an automotive ECU is a repetitive activity.

A bootloader software is designed to automate this flash reprogramming process and to handle the firmware upgrade. The bootloader program tests that the latest/updated version of the ECU program is usable at the device boot-up. If yes, then bootloader program installs and stores the newly installed firmware version before booting the device. Post this, the device boot-up is executed and device now runs in a fully protected environment on the latest version of the program.

At over 100 million lines of code in the total design, more than a conventional operating system, autonomous cars can no longer be viewed as pure mechanical devices. Increasingly, cars are linked and computer-like, with the potential to interact with cell phones, provide car passengers with the current weather and traffic alerts, and relay safety details to other automobiles and facilities surrounding them. Vulnerabilities in vehicle communications result in four obstacles to vehicle safety such as restricted access, restricted numerical capacity, volatile attack scenarios, and threats and essential danger to the life of drivers or passengers.

The main objective of this chapter is to develop a bootloader software using unified diagnostic protocols. The technique has to be implemented using any of the communication channel connecting the ECU. The bootloader should be able to provide a secure ECU reprogramming every time whenever a software update is requested. In order to mitigate the security treats faced by the connected vehicles, the software platform AGL has to be discussed. It should be able to provide a feasible update of software and application in an ADAS system. Apart from AGL, other platforms and their efficiency in terms of providing secure software update is detailed.

3.2 EXISTING SYSTEMS

An application to facilitate automotive remote update was developed (Zhang et al., 2018). The authors have discussed the related technology of bootloader software. As a means to upgrade the development software, the incremental change was brought in. They also introduced a comprehensive application architecture that involves memory model, state flow map, processes operating different applications, and applications frameworks. Generally, ECU programming began at first with the erasure of the whole memory area. Instead, the entire new codes were written into the ECU guide. The part of the codes that is modified is removed by itself and then revised. The physical space in flash is removed, and full writing is finished. It would minimize information fragmentations to maintain strong resource use. Using the gradual upgrade approach will significantly minimize the size of the software update and substantially reduce the installation period.

The steps needed to obtain a flash bootloader for an intelligent vehicle system were discussed (Bogdan et al., 2017) and also offer insights into the performance of using such an application. The methodology focused on the technology framework JCP2011, while the used hardware is a dashboard instrument with a RENESAS-produced microcontroller (3PD70F3537). The flash bootloader is an extension of the diagnostic feature and takes control of the dashboard instrument device upgrade using a unified diagnostic services (UDS) protocol. The diagnostic component of the program is described by a series of resources to be identified, checked, and performed by the device. The flash bootloader component of the diagnostic part accepts a limited set of commands from the entire protocol. This approach benefits from the sustainability point of view by reducing the need to travel to specialist centers to upgrade apps and also by downloading updates eliminating software bugs. It leads to more effective operation and smaller environmental and traffic effects.

Several cybersecurity challenges faced during the ECU update in vehicular communications were discussed by Rewinia et al. (2020). A three-layer framework (sensing, communication, and control) through which automotive security threats can be better understood was proposed. The sensing framework consists of vehicle dynamics and environmental sensors that are susceptible to attacks through eavesdropping, jamming, and spoofing. Attacks targeting the levels of sensing and communication will spread upward and impact the functionality, which can undermine the control layer's stability. The paper argues for a need for modern technologies to use systems of systems (SoS) approach to threats identification and mitigation. These architectures will concentrate on protecting sensitive units with cryptographic and noncrypto-based algorithms, registration, authentication procedures, and much more, such as power train ECUs.

An open-source secure software updates for Linux-based IVI systems were developed (Arthur Taylor, 2016). Cybersecurity and vehicle recalls are two significant factors in the production of automobile tech. Over-the-air upgrades are essential to enable car and system vendors to reduce protection and service threats as software is applied to automobile fleets, but there was no end-to-end open-source approach to handle upgrades until recently. Advanced telematic systems (ATS) has been collaborating with GENIVI and AGL in their development/reference networks to incorporate stable app updates. The paper provides an analysis of current open source OTA solutions, implemented the GENIVI SOTA open source solution, defined its design and security features, and defined the incorporation of the OTA system into the GENIVI development platform and the AGL framework.

Linux-based automotive security mechanisms could be used to make the connected cars safer and more secure in the future. The idea of virtualization is explored to establish the impression of several private subsystems or guest systems inside a single host system (Foll and Bollo, 2016). Each virtualization framework has frameworks for minimizing the usage of resources. Virtualization may be more or less complete, based on the model selected. Every computer has in certain cases a private kernel, with its own dedicated hardware resources. In other instances, virtual hosts share nearly everything: a single kernel, file system, RAM, I/O device with very limited separation, such as a private collection of libraries to curtail RAM or processor use.

Security hardening features based on ARM's TrustZone technology for infotainment systems is proposed that ensures confidentiality and integrity of critical applications (M. Hryhorenko, 2018). In addition, a strategy is introduced that enables mitigating the effect of certain attacks on the internal network of the vehicle. TrustZone is a hardware design, introduced with ARM processors on system-on-chip (SoC) computers and offers a protection mechanism to combat several vulnerabilities in embedded systems. TrustZone's primary function is partitioning both hardware and software resources into one of two worlds—safe, for protection subsystems and critical properties, and regular, for everything else. The planned research guarantees the safe elements remain secure and are essential.

The APPSTACLE project is discussed that simplifies the production phase of automotive systems by providing an accessible and safe software framework that links a wide range of automobiles to the software through in-car and internet connections (Pakanen et al., 2017). APPSTACLE project goal needs a wide spectrum of skills from networking systems to software-based technologies and services. The in-vehicle design tackles existing and lacking modules, utilities, protocols, APIs, and layers of software needed for stable and safe incorporation of the Car2X in-car framework. Direct cloud access standardization for in-car applications by a specific Car2X gateway interface also introduces the need for the prototypical specific Car2X gateway, which facilitates ECU wide remote access (e.g., through the gateway's CAN bus connectivity) to the cloud.

3.3 PROPOSED SYSTEM: UNIFIED DIAGNOSTIC SERVICES (UDS)

UDS is a diagnostic communication protocol (ISO 14229-1) in the vehicle electronics of the ECU, as laid down in ISO 14229-1. This is built from ISO 14230-3 (KWP2000) and ISO 15765-3 (Controller Area Network (DoCAN) Diagnostic Communication) (ISO 15765-2). This communication protocol is used in almost every new ECU made by OEM Tier 1 suppliers. The architecture of UDS protocol is shown in Figure 3.1.

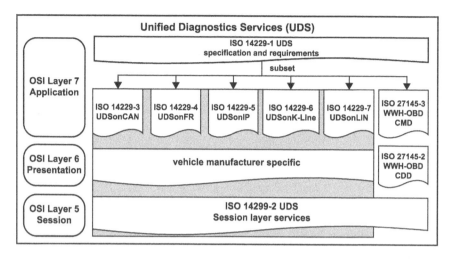

FIGURE 3.1 UDS protocol architecture.

The diagnostic device includes all control units mounted in a car, which are activated by UDS services. Unlike the CAN protocol, which uses just the Open Systems Interconnection (OSI) model's first and second layers, UDS systems are using the OSI model's fifth and seventh layers. The service ID (SID) and the services-related parameters are found in the 8 data bytes of a message frame received from the diagnostic device. Modern cars provide a software device for off-board diagnostics, allowing the vehicle's bus network to be linked to a computer (client) or software machine, which is called a tester. Therefore, the messages specified in UDS may be sent to the controllers who must provide the present UDS services. This helps the individual control units to query the fault memory or upgrade them with a new firmware.

The need for a standard diagnostic procedure was felt as OEMs combined/installed automotive ECUs and parts from various suppliers. This is because OEMs and vendors had to struggle with compatibility problems between different diagnostic protocols such as KWP 2000, ISO 15765, and over K-Line diagnostics compared to a single protocol. UDS is the recommended form of protocol for all diagnostic operations in off-board cars. Off-board diagnostics relates to testing the specifications of the car while the engine is in maintenance operation (whereas the car is stationary). ECU blinking and reprogramming may also be easily done with the help of a UDS stack. In comparison, the UDS protocol is very versatile and able to do more comprehensive diagnostics than other protocols such as OBD and J1939 (Yu and Luo, 2016). The most used services in UDS protocol are shown in the following Table 3.1.

3.4 FLASH BOOTLOADER

A flash bootloader module is built to upgrade the firmware without any specific equipment, including JTAG, being used. The flash bootloader module is the first program feature to be enabled during machine boot-up (after the device's power supply is turned on). The control is passed to the flash bootloader which searches for the

TABLE 3.1

Different Services Offered by UDS Protocol

Diagnostic Service Name	Description	Service ID Value (hex)	Sub-functions Supported?
Diagnostic session control	Controls the diagnostic session with the server	10	Yes
ECU reset	To reset the ECU	11	Yes
Security access	To unlock the secured ECU	27	Yes
Tester present	Indicates that server is still present	3E	Yes
Communication control	Controls the setting of communication parameters	28	Yes
Read data by identifier	Request data record value from the server	22	No
Write data by identifier	Write the data record values to the server	2E	No
Routine control	Start or stop a routine	31	Yes
Request download	Request for data download to the server	34	No
Transfer data	Download of data to the server	36	No
Request transfer exit	Request for download exit	37	No

firmware updated version. If a new version is available, flash bootloader can evaluate the upgrade to authenticate the source and test all predefined protection parameters for the program. If the authentication is effective, the bootloader module can write the latest update to the target address on the flash memory. Figure 3.2 depicts the flowchart of flash bootloader process.

3.5   UDS-BASED BOOTLOADER

UDS is the most effective protocol for applying bootloader, for ECU reprogramming purposes, according to the AUTOSAR Standard (Devi et al., 2018). Figure 3.3 shows the UDS based bootloader stack (Yu and Luo, 2016). Following are the important responsibilities of UDS in an ECU flashing operation:

- UDS sets the server into a programming session and start the flashing sequence.
- It handles the start and stop of the data transfer.
- UDS is responsible for the order and size of the data blocks to be transmitted or received.
- With the support of other software services, UDS helps the client to trigger a software reset event on the server.

3.6   SYSTEM IMPLEMENTATION: BASIC BOOTLOADER STACK

The development of bootloader software starts with the classification of requirements for the given application. The document consists of hundreds of requirements specific to both application and bootloader. For instance, requirements such as sensor

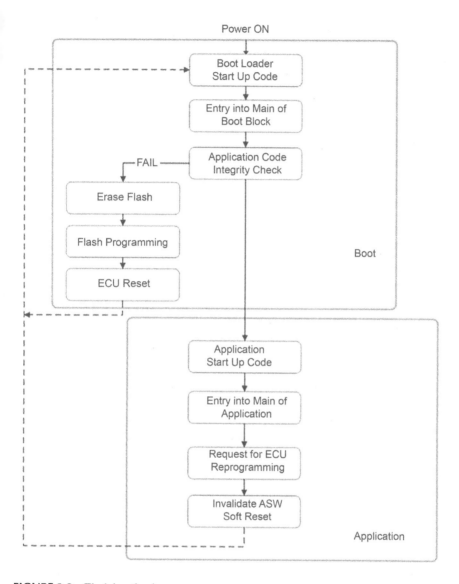

FIGURE 3.2 Flash bootloader process.

threshold level are considered for application alone, whereas the communication line bitrate, memory erase, etc., are specific to bootloader software. Thus, classification of requirements based on application and bootloader is mandatory.

Figure 3.4 shows the basic stack of components used in the design of bootloader software. The stack resembles the OSI model with the application layer containing the UDS protocol, and the physical layer consisting of the target microcontroller. The microcontroller abstraction layer consists (MCAL) of all the flash drivers used to erase and program the target microcontroller.

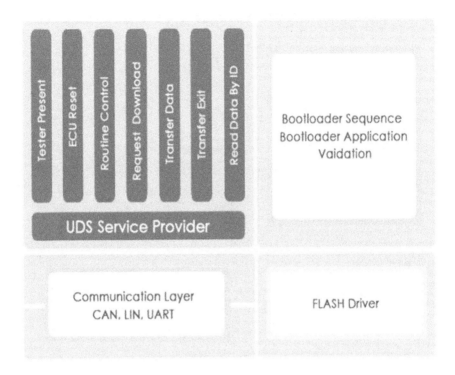

FIGURE 3.3   Block diagram for UDS-based bootloader.

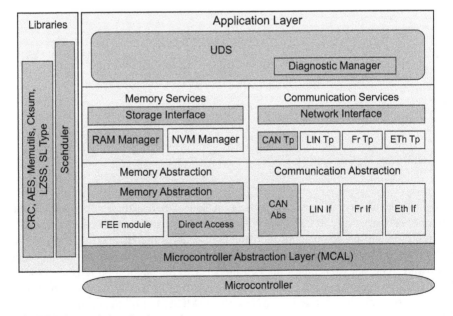

FIGURE 3.4   Basic bootloader stack.

The communication and memory layer is subdivided into abstraction and services stack, respectively. The communication layer contains different data transmission techniques such as CAN, LIN, Flexray, Ethernet, etc. With any one of the communication line, the software can be flashed into the microcontroller. The communication services layer provides a network interface (NetIF) component to interface with the UDS layer through the diagnostic manager. Nowadays, the diagnostic manager is replaced with the diagnostic communication manager (DCM) component which is AUTOSAR compliant (Sandhya Devi et al., 2019; Sivakumar et al., 2019). The memory layer consists of volatile and nonvolatile memory manager that can erase and write into corresponding memories of the target platform.

3.7 APPLICATION SOFTWARE FLASHING USING BOOTLOADER

Once the bootloader software is designed, the application is flashed into the given memory address in the target microcontroller. Every time the ECU is powered on or reset, the bootloader checks the validity of application, and programs it to the code flash

To program the ECU, the following sequence of services has to be sent as message frames in the communication line through UDS protocol. The services with their corresponding SID are shown below (UDS Protocol)

- Diagnostic Session control (0x10)
- Tester Present (0x3E)
- Communication Control (0x28)
- Security Access (0x27)
- Request Download (0x34)
- Transfer Data (0x36)
- Request Transfer Exit (0x37)
- ECU Reset (0x11)

3.7.1 HARDWARE REQUIREMENTS

The bootloader code can be checked using the microcontroller-connected debugger device. iSYSTEM debugger is the most commonly used method. iSYSTEM's latest entirely machine configurable iC5000 interface adapts with several specific processors and controllers as a multifunctional analyzer, production, and testing tool.

The winIDEA integration programming environment (IDE) offers the visual perspectives required for debugging the embedded application. WinIDEA includes much of the normal features of an IDE at the simplest stage, such as breakpoints, walking, and application programming. WinIDEA can also simulate the application's timing and code coverage using the trace interface while assisted by a target microcontroller, as well as blend data collected by the IOM6 accessories.

CANoe analyzer is a Vector Informatik GmbH's production and testing analysis tool. The device is mainly used by car producers and suppliers of the ECU to build, evaluate, model, check, diagnose, and start up ECU networks and individual ECU's. It is particularly well suited for ECU production of traditional vehicles as well as hybrid vehicles and electric vehicles due to its extensive use and large number of assisted vehicle bus systems.

3.8   SIMULATION

The programming part is carried out by the following steps:

1. Connect the iSYSTEM debugger to the ECU through its debugging port. With the help of winIDEA IDE, as shown in Figure 3.5, the bootloader software is tested and analyzed for errors.
2. The CANoe analyzer is connected to the CAN communication port in the ECU. The communication technology chosen here is the CAN protocol.
3. In the CANoe tool, the respective CAN channel is selected for data transmission as shown in Figure 3.6.
4. The message frames are created and sent through the CAN bus to verify proper communication from the ECU.
5. For every diagnostic request, a response message frame is received from the ECU.
6. Once the bootloader software is completely verified, the given application is flashed into the ECU memory through flashing tool.
7. In the flashing tool, the start and end flashing address of the memory has to be selected.
8. The diagnostic services sequence has to be properly given in the flashing tool so that the flashing of application software takes place correctly.

Once the parameters are configured, the download button is clicked to program the target microcontroller.

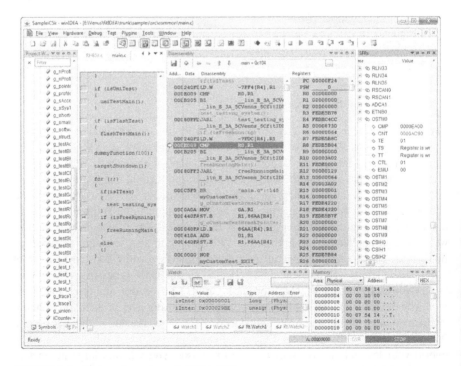

FIGURE 3.5   Debugging using winIDEA IDE.

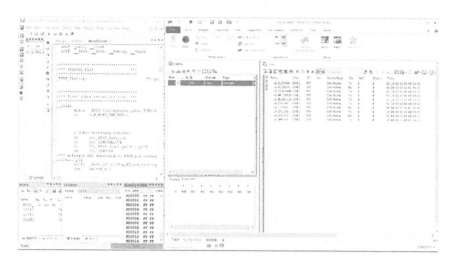

FIGURE 3.6 CANoe tool for transmitting and receiving message frames.

3.9 SECURE ECU SOFTWARE UPDATE: SOFTWARE INTEGRITY

The safety and security of the customer is an important aspect in designing an automobile software. If a hacker tries to bypass some of the device or computer code, then the machine's full security is breached. As in connected vehicles, which is the immediate future, only the initial applications accepted for the boot will be accessible. Instead, no new details will be applied to the application before updating the program. Thus, an appropriate program needs to be uploaded to the ECU.

The bootloader shall check the integrity of the application software during the booting process (Kargl et al., 2008). To validate the code, it should use some method of cryptography (Zhang et al., 2018). This strategy is known as trusted boot or secure boot. Ensuring correct computer execution requires extensive cryptographic computation. While such operations without a crypto-coprocessor may have reduced enduser experience to a barely appropriate level in the past, modern hardware like ARM-V8 with cryptographic extension does not face such limitations today.

After the completion of booting process, the software will have to handle various applications. Many applications with an independent life-cycle can also be provided and maintained by third parties. Thus, before accepting an application, the system should

- Check the integrity of the application code.
- Handle the dependencies in the system.
- Install and start the application.

With hundreds of sensors, actuators, and ECUs that need to work together while maintaining optimal efficiency, protection, and security, the cryptographic methods alone are insufficient to achieve the objectives. Therefore, it is important to adopt a far more effective approach.

3.10   AUTOMOTIVE VIRTUALIZATION

Virtualization is a methodology associated with the separation or partitioning of resources to build several virtual execution environments. The aim of virtualization is to establish the impression of numerous private subsystems or guest systems within a single host system. Each virtualization framework has frameworks for minimizing the usage of resources (CPU, Ram, Filesystem). Every virtual machine runs a private operating system in hardware virtualization, utilizing hardware separation inside the CPU, under the control of limited software called a "hypervisor." The future application in the automobile industry could be to hold the master ECU separated from the IVI by utilizing a single hardware unit. The virtualization of the software, often called containerization, is the use of a Linux system called unsharring. This virtualization also has an expense, but stays very small as opposed to virtualization of the hardware. It may be used in automotive to separate untrustworthy systems from the baseline utilities.

3.10.1   AGL VIRTUALIZATION CONCEPT

Automotive Grade Linux is developing a Linux-based, open-source computing framework to function as the de-facto automotive industry norm (Sivakumar et al., 2020; Devi et al., 2019). The AGL virtualization solution seeks to include a virtualization infrastructure that can be used to integrate multiple automotive functions in a single hardware framework, as it is or expanded (Contini, 2014). AGL is not developing new hypervisors, but is leveraging existing open-source solutions to consider them as modules for its architecture. The concept of virtualization in AGL is shown in Figure 3.7. The AGL application architecture already supports separation of programs based on namespaces, Cgroups, and simplified compulsory access control module (SMACK), which relies on security attributes of files or processes that are tested by the Linux kernel each time an action is processed and also work with the stable boot techniques. However, where several applications with specific security and safety criteria (infotainment, instrument cluster, telematics, etc.) need to be installed in the network, the control of these security features is complicated and an additional degree of separation is needed to better separate such applications from one another. Thus, the AGL virtualization architecture helps boost device stability and separate specific programs from the AGL network, but often from developers from third parties. Other security protocols that require specific hardware for safe upgrading of applications are listed below.

- Crypto co-processor: Secure communications (SSL/TLS) or encryption of disks involve encryption such as Advanced Encryption Standard (AES). Using built-in processor instructions or coprocessors, both of these cryptographic functions are very accelerated.

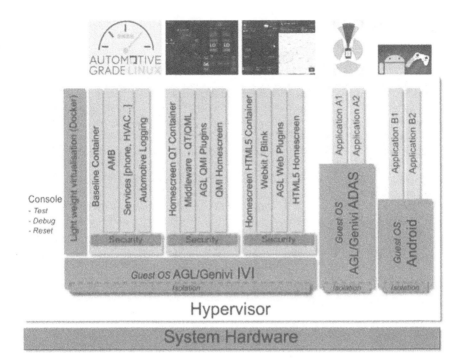

FIGURE 3.7 AGL virtualization concept.

- ARM TrustedZone: This ARM function divides the processor and other peripherals into two zones: one trusted and one nontrusted. Secrets will instead be kept inside the zone of confidence to prevent exposure. This profoundly embedded technology within SoC architecture allows it effective to hold the hardware hidden.
- Secure RAM: Any passwords stored before use will be decrypted. Nevertheless, decrypted keys on every hardware bus should be kept invisible. The SoC provides protected RAM inside the trusted zone to accomplish that.
- UICC/SIM cards: SIM cards are meant to keep secrets and may be used to this purpose.

3.11 CONCLUSION

The bootloader software for ADAS system was designed using the basic bootloader stack. Different components are integrated and tested with the help of winIDE. The CANoe tool is used to analyze the flow of request and response messages through the CAN communication line. The application software is flashed into the ECUs memory using the flashing tool. The security challenges faced during the ECU software update have been discussed. AGL security mechanism has been proposed to provide a secure application installation after booting of the system. Thus, with the help of the bootloader software, we can check the validity of the application and provide a safe software update.

REFERENCES

D. Bogdan, R. Bogdan, and M. Popa, 2017. Design and implementation of a bootloader in the context of intelligent vehicle systems, IEEE Conference on Technologies for Sustainability (SusTech), 2017, pp. 1–5.

A. Contini, 2014. Melissa Logan. JVC Kenwood, Linaro, and Opensynergy Join Automotive Grade Linux. Automotive Grade Linux (AGL). [Online] Available https://www.automotivelinux.org/news/announcement/2014/11/jvc-kenwood-linaro-and-opensynergy-join-automotive-grade-linux Accessed on 2020-06-12.

F. A. Foll and J. Bollo, 2016 "Linux Automotive Security-Safer and more Secure", IOT BZH, Version 1.0.

M. Hryhorenko, 2018. "Towards Enhanced Security for Automotive Operating Systems", Florida Institute of Technology, Melbourne, FL.

R. S. Devi, P. Sivakumar, and R. Balaji, 2018. AUTOSAR Architecture Based Kernel Development for Automotive Application. In International Conference on Intelligent Data Communication Technologies and Internet of Things (pp. 911–919).

ISO 14229-1, Road vehicles—Unified diagnostic services (UDS).

ISO 15765-2, Road vehicles—Diagnostic communication over Controller Area Network (DoCAN).

F. Kargl, P. Papadimitratos, L. Buttyan, M. Müter, E. Schoch, B. Wiedersheim, T.-V. Thong, G. Calandriello, A. Held, A. Kung, and J.-P. Hubaux, 2008 "Secure vehicular communication systems: Implementation, performance, and research challenges", IEEE Communications Magazine, 46(11):110–118.

P. Sivakumar, R. S. Sandhya Devi, A. Neeraja Lakshmi, A. Vinoth Kumar, and B. Vinod, 2020. "Automotive Grade Linux Software Architecture for Automotive Infotainment System", 5th International Conference on Inventive Computation Technologies (ICICT-2020).

P. Sivakumar, R. S. Sandhya Devi, and B. Vinoth Kumar, 2019. "Analysis of Software Reusablity Concepts used in Automotive Software Development using Model-Based Design and Testing Tools". Alliance International Conference on Artificial Intelligence and Machine Learning (AICAAM), pp 79–89.

Olli-Pekka Pakanen, A. Banijamali, A. Haghi Ghatkhah, O. Liinamaa, G. Destino, P. Kuvaja, M. Latva-Aho, M. Oivo, M. Frisk, and Z. Laaroussi, 2017. "APPSTACLE-Breaking the Silos in Automotive Software and Systems Development". European Conference on Network and Communication.

R.S. Sandhya Devi, P. Sivakumar, and R. Balaji, 2019. AUTOSAR Architecture Based Kernel Development for Automotive Application. In HemanthJ., FernandoX., LafataP., BaigZ. (eds) International Conference on Intelligent Data Communication Technologies and Internet of Things (ICICI) 2018. ICICI 2018. Lecture Notes on Data Engineering and Communications Technologies, vol 26. Springer, Cham.

R.S. Sandhya Devi, P. Sivakumar, and M. Sukanya, 2018. Offline analysis of sensor can protocol logs without can/vector tool usage. *International Journal of Innovative Technology and Exploring Engineering*, S2:225–229.

Z. Rewinia, K. Sadatsharana, D. FloraSelvaraja, S.J. Plathottamb, and P. Ranganathana, 2020 "Cybersecurity challenges in vehicular communications", *Vehicular Communications*, 23: 100214.

A. Taylor, 2016. Open-Source secure software updates for Linux-based IVI systems. Advanced Telematic Systems.

UDS Protocol. [Online]. Available: https://embedclogic.com/uds-protocol/

J. Yu and F. Luo, 2016. "Research on Automotive UDS Diagnostic Protocol Stack Test System", *Journal of Automation and Control Engineering*, 4(5):388–392.

J. Zhang, X. Zhu, and Y. Peng, 2018. Implementation and research of bootloader for automobile ECU remote incremental update, AASRI International Conference on Industrial Electronics and Applications, Clean Energy Automotive Engineering Center of Tongji University, Shanghai, China.

4 Advanced System Requirements for Automotive Automation

H. Suneeta, M. Manohar, and S. Harlapur
Department of ECE, Vemana I T, Bangalore, India

CONTENTS

DOI: 10.1201/9781003269908-4

SHORT SUMMARY

Technology has ensured in making car driving experience safer and comfortable. As the days progress, the advancement in automotive technology is an ideal collaboration between man and machine. All this is possible with technology spreading its wings in a rapid rate. The benefit of utilizing nanocomposites in outer parts and drive extravagance vehicles positively affect market demand. This development is possible only by combining modern technologies with human thirst for excellence, knowledge, and step up the every ladder of science and even the wars that provided us the tools have now been used in cars like antennas, radars and wireless communication to ensure a safer driving experience.

4.1   INTRODUCTION

Today, the car world faces preferably more changes over the previous decades. In contact to the past, we mostly saw developmental changes—better arrangements in territories of wellbeing, solace, or driver help. In pace with computerized advances, increasingly more hardware entered the vehicle prompting around 100 electronic control units (ECUs) in current premium vehicles, a few hundred sensors and actuators, many correspondence transports with a huge number of different signs. Center standards, nonetheless, kept unaltered; like ignition motor, and in this manner, the vehicle driven by someone's driver who is associated with their condition mostly by human's detects (aside from radio or potentially phone). Having an own vehicle could be a sensibly superficial point of interest, and vehicles are created by customary car OEMs. Presently, these ideal models (incompletely) aren't any more drawn out substantially and that we face increasingly more problematic change, as: Traditional ignition motors are supplanted or enhanced by electric driving. Explanations behind this are both becoming natural understanding and regulative powers. Nearby outflow free driving is distinguished as a technique to bring down fine residue which turns out to be increasingly more applicable gratitude to expanding urbanization and builds autonomy from fuel.

New styles of organizations are entering in the car market like Apple, Google, Tesla, etc. Accordingly, Mercedes-Benz railcar has characterized four vital point territories where they intend to assume a main job. They're connected, autonomous, shared and service, and electric drive, condensed as CASE (Swetha, S., & Sivakumar, P. 2021). A center of availability offers a driver admittance to their vehicle through an application or web content. Capacities like network-based stopping, sharing data on free parking areas gathered by leaving sensors consolidate Mercedes-Benz vehicles, plan to share data between vehicles. In the previous years, there was no such major increment of capacities on the gratitude to independent driving that is accessible in research vehicles, yet available to ordinary drivers. A framework called "drive pilot" offers semi-mechanized driving on roadways, underpins and surpassing which may create a crisis like slowdown.

In the car business, particularly inside the top-of-the-line market, the usefulness of electronic segments is turning out to be increasingly more intricate at a truly quick rate. Around 33% of improvement costs is spent for electric/electronic advancement these days, and this sum stays expanding. At the indistinguishable time, numerous parts are created, incorporated, and tried over a succession of prototyping stages. Besides, parts are created in a numerous variation and at various timetables. As an outcome, the detail exercises have arrived at level of multifaceted nature that surpasses the limits of what are frequently sensibly upheld by regular content preparing frameworks utilized by some neighborhood saints, in light of the fact that these frameworks don't adequately uphold the executives and following usefulness. In light of these reasons, necessities building cycles, strategies, and devices are being steered at Daimler Chrysler in different improvement ventures throughout the utmost recent years.

4.2 REVOLUTION OF AUTOMOTIVE ENGINEERING

Imagine yourself not long from now driving your new vehicle along a provincial interstate roadway on an excursion for work. The journey control is keeping up the speed at a consistent 100 km/hr (62 mph) and there is generally little traffic. As you approach a slower vehicle, the speed-control framework eases back your vehicle to coordinate the speed of the slower vehicle and keep up a protected separation of around 53 m (165 ft) behind the slower vehicle. When approaching traffic clears, you enter the passing path and your vehicle naturally speeds up as you pass the slower vehicle (Robert Bosch Gmbh 2013).

You press a catch on the guiding segment and a picture of a guide shows up faintly obvious on the windshield before you. This guide shows your current position and the situation of the objective city. You are chatting on your phone to your office about certain adjustments in an agreement that you would like to arrange. After the directions for the agreement changes are finished, a printer in your vehicle produces a duplicate of the most recent on parcel adaptation.

The on-board theatre setup is playing music for you at an agreeable level comparative with the low-level breeze and street commotion in the vehicle. Subsequent to finishing your telephone discussion, you press another catch on the directing haggle music is supplanted by a recorded exercise in English action word formation, which you have been examining.

"You have fuel staying for another 50 miles at the current speed. Your objective is 23 miles away. Suggest refueling in the wake of leaving the expressway. There is a station that acknowledges your electronic credit close to the leave (you know, obviously, that the electronic credit is actuated by embedding's the fuel spout into the vehicle). Additionally, the left back tire pressure is low and the motor control framework reports that the mass wind current sensor is irregularly failing and ought to be adjusted soon."

Then you embed the right plate in the navigation CD player as mentioned and the guide show on the windshield changes. The new showcase shows a nitty gritty guide of your current position and the course to your objective. As you approach as far as possible, the vehicle speed is naturally diminished to the lawful furthest reaches of 55 mph. The voice message framework talks once more: "Leave the thruway at exit 203, which is one-half mile away. Continue along Austin Road to the subsequent crossing point, which is Meyer Road. Turn right and continue 0.1 mile. Your objective is on the right-hand roadside. Remember to refuel."

This situation isn't as fantastical as it sounds. The entirety of the occasions portrayed is actually conceivable. Some have even been tried tentatively. The electronic innovation needed to build up a vehicle with the highlights depicted exists today. The genuine execution of such electronic highlights will rely upon the expense of the hardware and the market acknowledgment of the highlights.

4.2.1 Utilization of Electronics in Automotive Engineering

Microelectronics will give many energizing new highlights to cars. Hardware has been moderately delayed in going to the car fundamentally as a result of the connection between the additional expense and the advantages. Truly, the main hardware (other than radio) was brought into the business car during the last part of the 1950s and mid-1960s. Nonetheless, these highlights were not generally welcomed by clients, so they were ceased from creation autos.

Ecological guidelines and an expanded requirement for economy have brought about hardware being utilized inside various car frameworks. Two significant occasions happened during the 1970s that began the pattern toward the utilization of current gadgets in the vehicle:

- The presentation of government guidelines for exhaust emanations and mileage, which required preferable control of the motor over was conceivable with the techniques being utilized; and
- The advancement of moderately ease per work strong state computerized hardware that could be utilized for motor control.

Hardware are being utilized now in the vehicle and likely will be utilized much more later on. A portion of the present and expected applications for gadgets are

- Electronic motor control for limiting fumes emanations and amplifying mileage.
- Instrumentation for estimating vehicle execution boundaries and for analysis of on-board framework glitches.

- Driveline control.
- Vehicle movement control.
- Security and accommodation.
- Diversion/correspondence/route.

4.3 AUTOMOTIVE REQUIREMENTS ENGINEERING

Necessities to build the advancement in the car functionality will raise the difficulty in design or occasions when there are barely any individuals and assets included. However, when either the quantity of individuals or the necessary assets develop, arranging can immediately get troublesome and execution significantly more so. Hence, compelling joint effort and correspondence gets basic. In like manner, effective item improvement relies enormously upon coordinated effort among the numerous people and groups included, including frameworks, programming, electrical and mechanical architects, and others, for example, the designing chief. They should all comprehend and work toward the undertaking's prerequisites, which is the string that interfaces all task partners and groups with regular goals.

Prerequisites additionally make a connection between improvement groups and clients, providers and colleagues. At last, necessities characterize an item intended to fulfil a client need or commercial centered interest—so they should tune into those requirements and requests. For modern item advancement, the necessities designing cycle assists organizations with overseeing complex prerequisites, improve group cooperation, and at last produce excellent items more expense viably. Here we discuss about the accepted procedures and advantages of prerequisites building, and how car producers can defeat the present quality and cost difficulties utilizing those prescribed procedures upheld by IBM arrangements. A few contextual investigations delineate how car unique hardware producers (OEMs) and industry providers have utilized IBM Rational® DOORS® programming to help improve correspondence and joint effort through the prerequisites building measure (Sivakumar, P. et al 2015). Thus, these associations can help efficiency, increment time and cost investment funds, and make better finished results.

Meeting difficulties in the car business, due to the poor financial atmosphere, the issues confronting the car business pervade the everyday lives of industry experts and purchasers as well. The title of everyday papers caution of expanding fuel expenses that change purchaser mind. Furthermore, TV report tickers give consistent tokens of the strain to decrease CO_2 outflows and the change in return rates. With such worries at the exploiting edge, car organizations must zero in on better approaches to create more excellent items quickly and less expense viably.

Generally significant, they should obviously comprehend what they are working all through the improvement lifecycle as such; they have to know whether all client necessities are satisfied. Hence, car organizations can effectively oversee item advancement and designing by adopting a prerequisite driven strategy. Prerequisite's building can help to meet a portion of the accompanying key difficulties. Practical development OEMs and providers need inventive items to win market share. In any case, to improve net revenues, they have to distinct their brands and upgrade execution while quickening lifecycles, lessening costs, guaranteeing quality, and conveying item greatness.

A prerequisite driven way to deal with item improvement and designing can assist automakers with creating models more expensive, quicker, and at a higher caliber which is necessary for the car business. The features like supplier collaboration reducing advancement cost and overseeing unpredictability require close affiliation and incorporation among OEMs and providers and some key changes in the manner they work together. Providers need to adjust plan and item improvement to quickly changing OEM prerequisites while expanding item quality and diminishing chance to market and cost. Expanded interest for electronic and programming content in vehicles, The intricacy of the vehicle will develop exponentially as future advancements empower it to turn out to be heavier and more associated. Heads we talked with gauge that 90 percent of future advancement will be founded on hardware, the majority of which will be installed software (B.L. Krovitz, 1998).

As the pace of advancement and multifaceted nature of these implanted gadgets increase, car makers will require a more taught way to deal with frameworks and programming improvements. In what capacity can car organizations deal with these difficulties? An initial step is to wipe out helpless necessities rehearses and embrace a prerequisite designing cycle for item improvement. Characterizing prerequisites designing requirements building as far as frameworks and programming building characterizes, oversees, and efficiently tests necessities for a framework. It does as such in three phases: needs examination, necessitates investigation, and prerequisites determinations.

In spite of the fact that this meaning of necessities building is over ten years old, a standard cycle has as of late advanced with the accessibility of incorporated setups of mechanized lifecycle improvement devices highlighting prerequisites of the executive's arrangements. In essential terms, necessities designing aide's OEMs comprehend what they plan to work in two phases. The primary stage is to characterize prerequisites in advance. The second is to oversee them by having clear perceivability all through the item lifecycle. The main phase of prerequisites building necessities definition comprises of four sections: disclosure, examination, determination, and check. Prerequisites are the executives at the subsequent stage, disentangles and upgrades correspondence, and joint effort among all groups and partners, bringing about better necessities the board all through the association. This stage empowers architects to:

- Evaluate the impact of proposed changes.
- Trace singular necessities to downstream work items.
- Track prerequisites status during improvement. Therefore, they can screen venture status by realizing what level of the designated necessities have been either:
  - Implemented and checked.
  - Just executed.
  - Not yet completely executed.

Necessities designing aide's OEMs comprehend what they mean to work by first characterizing prerequisites and afterward overseeing them all through the item lifecycle. At the same time, necessities associate the worldwide building groups frameworks, programming, electrical/electronic and mechanical, and keep them all the more distinctly centered around regular destinations. Besides, prerequisites give a fundamental association between the building groups and other fringe partners, including providers, clients, and inward lawful and quality-affirmation groups. Utilizing a prerequisite building system and a supporting apparatus for necessities the board and recognizability, architects can nicely tailor advancement practices to suit the task type, limitations, and hierarchical culture. Prerequisites designing for the car business automotive frameworks and subsystems have many interlinked, subordinate parts, so car specialists and engineers must see how all the parts cooperate. To do as such, they should have the option to see the entirety of the associations. Something else, the final product will be twisted and not what was normal by the client.

As the interest for on-board programming expands, these multifaceted frameworks become much more unpredictable. Hence, building groups must fill in as a strong unit to oversee changes and reuse segments so they can react to client requests. Furthermore, necessities and recognizability the executives programming, for example, rational DOORS, can empower those building groups to oversee framework multifaceted nature, upgrade correspondence and coordinated effort, improve item quality, and follow industry norms. Necessities and recognizability the board programming give building groups more prominent authority over overseeing and examining a huge number of prerequisites for car items like:

- Decompose beginning client necessities into definite prerequisites.
- Link prerequisites and configuration to confirm that necessities are fulfilled by the structure.
- Trace conditions among necessities and changes.
- Analyse the effect of necessities changes.

Designing groups utilizing rational DOORS have more prominent command over overseeing and investigating a huge number of necessities for car items. By utilizing these robotized prerequisites, the executive's device as the foundation of the necessities designing condition, they can decrease advancement time and increment profitability through normalized measures. Utilizing the recognizability usefulness of rational DOORS, architects can follow an enormous volume of highlights back to the prerequisites. They would then be able to reuse the necessities for regular parts over numerous product offerings and models. The groups gain efficiency, while the organization sets aside cash and increases quicker commercial center conveyance of client driven highlights. Improve correspondence across the world with better correspondence and coordinated effort among universally different designing groups including frameworks, programming, electrical, electronic and mechanical, and different partners, for example, clients, providers, and inward quality affirmation groups is a vital aspect of the necessities building measure.

Since prerequisites are partaken in the rational DOORS focal storehouse, geologically scattered groups can effectively share data and work together more successfully just as invest less energy. Accordingly, they can get particulars directly toward the start of the task since they are working from a characterized set of necessities. Balanced DOORS underpins a necessary designing way to deal with prerequisites the executives, assisting with explaining arrangements, and exchanges among OEMs and providers. Prerequisite designing is required for the car business. Better correspondence explains concession to and arrangement of:

- The necessities that the provider will fulfil.
- The manner by which the provider plans to react to the necessities, for example, the proposed items.
- The manner by which created items will be confirmed on conveyance.
- Evidence that the conveyed items are palatable.

Manufacture top notch frameworks system quality is crucial to security and vehicle execution. By utilizing rational DOORS in a prerequisite designing methodology, every necessity is connected with testing to approve its presentation. Groups can likewise incorporate prerequisites and approve them against a model, making it simpler to discover holes among necessities and models being developed. Thus, frameworks designers can accomplish the objective of giving top notch, safe items that really address clients' issues. Agree to principles requirements designing engages car organizations to oversee consistence. They can utilize the prerequisites detectability in rational DOORS to report the connections among measures and the consistence structure. Anytime in building the lifecycle, groups can without much of a stretch check for prerequisites that are not fulfilled by the plan or for structure components with no connected necessities. They can likewise present car consistence measures, for example, automotive open system architecture (AUTOSAR) (Sandhya, D. R., Sivakumar, P., & Balaji, R. 2019) and the Software Process Improvement and Capability dEtermination (SPICE) model as a component of the necessities recognizability measure in Rational DOORS.

Discerning DOORS are examples of overcoming adversity in the car business. Many worldwide car organizations and their providers have received necessities designing upheld by IBM answers for effectiveness and cost, successfully gain group profitability and greater items just as quicker commercial center conveyance.

Utilizing rational DOORS, they have the ability to oversee and examine in excess of 100,000 prerequisites in complex tasks intended to assemble the inventive vehicles that drive net revenues.

Judicious DOORS can help improve prerequisites perceivability all through the building lifecycle. The detectability of rational DOORS abilities additionally assists groups with guaranteeing that basic highlights are not missed. European OEM improves necessities approval OEMs and providers need to approve prerequisites against models to expand model quality and use discernibility to guarantee that basic prerequisites are not missed from the advancement cycle. The European OEM utilizes rational DOORS to oversee and approve necessities against creating models and to distinguish holes among partner and framework prerequisites. The improvement groups can remain centered all through the whole cycle of making a framework, which is especially significant when they need to create complex frameworks over a few unique brands. Thus, the OEM has decreased time spent dissecting necessities for irregularities, including missed lawful prerequisites. OEM streamlines provider joint effort communication among various topographically scattered groups and the connection among OEMs and providers can influence the perfection cycle and the nature of new items.

A main European OEM and a main German provider utilize rational DOORS as a focal help instrument for their framework's designing cycle. The OEM likewise utilizes rational DOORS to trade necessities information with its providers at the electrical/part level. It has upgraded correspondence and coordinated effort inside complex tasks and limited dangers, including revamps, reviews, combination issues, missing necessities, and missed cutoff times. Besides, it can approve necessities at each progression and guarantee that it doesn't miss basic data.

4.4 COMMUNICATION AMONG DEVICES IN AUTOMOTIVE ENGINEERING

Clever transportation system incorporates eight wide regions of utilizations, for example, advanced traffic information system, advanced traffic management systems, automatic vehicle control system, commercial vehicle operations, advanced public transport systems, emergency, infotainment expectations, and comfort to drivers and passengers, safety, electronic payments, and so on. Security, mobility, and comfort are the three fundamental explanations behind expanding the contribution of gadgets in vehicles and, thus, in intelligent transportation framework.

Vehicular correspondence guarantees traffic the executives, effective, and simpler support, more fun and infotainment which at last lead to more secure streets, upgraded comfort for driver and travelers (safe transportation). Completely the vehicular communication can be delegated.

4.4.1 VEHICLE TO INFRASTRUCTURE (V2I) AND INFRASTRUCTURE TO VEHICLE (I2V)

In thruway building, improving the security of a street can upgrade generally speaking effectiveness. Vehicle infrastructure correspondence targets upgrades in both security and effectiveness (Subburaj, S. D. R. et al 2021). It is remote correspondence cooked by vehicular ad-hoc networks, which gives correspondence among vehicles and close by fixed equipment's, generally portrayed as side of the road gear's or framework.

4.4.2 INTER-VEHICULAR (V2V)

Entomb vehicle correspondence is administered by remote correspondence and is cooked by mobile ad-hoc networks (MANETS) and vehicular ad-hoc networks (VANETS). A VANET, is a type of mobile impromptu system, to give interchanges among close by vehicles.

4.4.3 INTRA-VEHICULAR

Intra-vehicular correspondence depicts as trade of information inside the ECUs of the vehicle, which are associated with vehicular applications. Major intra-vehicular correspondence is of wired sort, for example, based on arrangement. There are a few applications wherein remote Intra-vehicular correspondence is accounted.

4.5 COMMUNICATION AMONG DEVICES

One can envision the complications in question and dynamic headways with around 8–10 ECUs in vehicle in mid-90's through around 50 microcontrollers toward the start of this decade, for example, 2000. Today, in very good quality vehicles, it is entirely expected to have around 100 ECUs trading up to 2500 signs between them. Car systems eradicates heavy wiring binds, upsurges vehicle security and reliability, future development effectively conceivable, consistency and tuning in activities, fast, reliable and efficient intercommunication, organizes the data stream, assurance of dedicated correspondence.

A PORTION OF THE AUTOMOTIVE NETWORK CONTROLLERS

4.5.1 LOCAL INTERCONNECT NETWORK (LIN)

The LIN is a sequential correspondence convention appropriate for systems administration sensors, actuators, and different hubs continuous frameworks. It is 12V single wire, universal nonconcurrent beneficiary transmitter (UART), single-master, multiple-slave (up to 16 hubs), up to 20 kBits/s information rate car arrange convention. LIN bus application models:

- Steering wheel: Cruise control, wiper, turning light, climate control
- Seat: Seat position motors, occupant sensors, seat control switch
- Door: Mirror switch, central ECU, power window lift, door lock
- Safety and investigation: Automotive black box.
- Controller area network (CAN)

The CAN transport is utilized in vehicles to associate motor control unit and transmission, or to interface the entryway locks, atmosphere control, seat control, and so on (Devi, R. S., Sivakumar, P., & Sukanya, M. 2018). It is asynchronous multiace correspondence convention sequential transport having information pace of up to 1 Mbps. Examples of CAN bus application:

- Safety power train: Electronic parking brake, vacuum leak detection automotive black box.
- Chassis: Watchdog, motor control, electronic throttle control body control: low-end body regulator (lighting, network communication) power door, power sunroof, power lift gate.
- Flex-ray organize.

Flex-beam organizing norms take a shot at the rule of TDMA and have dual-channel architecture. It has host processor which controls the correspondence cycle by means of automotive networks based intra-vehicular communication applications 209 correspondence regulator and transport driver (Bajaj, P., & Khanapurkar, M. 2012). EACH flex beam hub has two physical channels A and B encouraging information pace of up to 10 Mbps through per channel. It tends to be utilized as system spine with CAN and LIN. Examples of flex-ray network bus application:

- Data backbone: For different transports (LIN, CAN, and MOST).
- Safety-basic applications: X-by-wire, automotive black box.
- Distributed control framework applications: Power train applications.
- Chassis applications requiring calculations across different ECUs.
- Media arranged framework transport (MOST).

MOST is a rapid interactive media that organize innovation ordinarily incorporates up to 64 MOST gadgets having plug n play activities. MOST can work up to 15 uncompressed sound system channels or up to 15 MPEG1 channels for sound video correspondence.

4.5.2 Steps to Plan Intra-Vehicular Correspondence Applications

- Depending upon the sort of intra-vehicular communication applications, organize regulator and host processor are to be chosen.
- The intra-vehicular correspondence application can be homogenous car regulator based or it very well may be of heterogeneous kind.
- Design and change (subsequent to testing emphases) of interface design model (with appropriate displaying device/language like VHDL, LABVIEW, and MATLAB and so forth) of system regulators.
- Downloading the model document made with configuration device in to appropriate programmable stage (like FPGA, CPLD and so on) and result validation.
- Interfacing with electromechanical frameworks for reproduction of ongoing information from sensors, actuators, and testing for activities.
- Installation and commissioning.
- Testing the structured framework progressive condition.

4.6 NANOMATERIALS COATING ON AUTOMOTIVE SYSTEMS

The role of the coatings in automotive industry is mainly to avoid corrosion and make the surface as corrosion resistant. Here chemical degradation or corrosion happens due to the exposure of automotive surface to UV light, or moisture and more. The automotive coatings are surface coatings carried out during finishing processes. The functions of automotive coatings are not only limited for corrosion resistant or decorative purpose, but many. These coatings make automotive more attractive by giving different colors. Along with these, even we can add properties like, smoothness, optical reflectivity, heat resistance, scratch resistant, and so on. So, it is a surface phenomenon undergoes by adsorption process.

The study is to emphasize and present in brief the prospective impact of nanotechnology on automotive industry to enhance the fabrication of latest automotive designs with an optimized wellbeing performance along with the impact on environment.

Especially the nanomaterials for locomotive solicitations are envisioned to enhance the drop in engines discharges and body weight, for safe drive, soundproof vehicles, self-mending body, and windscreens (Presting, H., & König, U. 2003).

Nanocomposites are new class of polymeric/non polymeric filled composites with remarkable mechanical, physical, and handling properties made effortlessly. When the composite material has reinforcement in which at least one dimension in the nanometer scale or less than 100 nm, then the case of nanocomposites. Therefore, the nanocomposites are composites that consist of the base material and the nano size of the reinforcement which do not dissolve in one another. The viability of the nanoparticles is with the end goal that the measure of material included is ordinarily just 0.5–5% by weight (Coelho, M. C., Torrão, G., & Emami, N. 2012).

However, the subsequent materials have properties, which are better than traditional microscale composites.

The utilization of nanocomposites in vehicle parts and frameworks is required to improve fabricating speed, upgrade natural and warm dependability, advance reusing, and diminish weight. Use of this innovation is simply to noncritical auxiliary parts like front and back parts, cowl vent flame broils, valve/timing spreads, and truck beds; however it could decrease a few billion kilograms of weight for each year. In addition, use of permeable nanocomposites can be utilized as toxic filters, which mechanically or by catalytic reaction reduces a release of soot particles or toxic gases (Lin, Y. C. et al 2008).

The car business benefit from nanomaterials in many domains like, casings and body parts, motors and powertrain, tribological perspectives, paints and coatings, suspension and breaking frameworks, oil, tires, fumes frameworks and exhaust systems, and electric and electronic gear, although decreasing vehicle weight, improving material capacities, expanding solace degree and flexibility, raising cost efficiency. Practically, all the car segments can be improved by nanotechnology as represented (Coelho, M. C., Torrão, G., & Emami, N. 2012) in Figure 4.1.

Nano-grains of glasslike stage have potential in tackling a few issues of regular coatings. These coatings give high hardness, low outstanding pressure, and high durability and show generally excellent tribological properties. Higher coefficient of warm extension (CTE), hysteresis and lower warm diffusivity, higher hardness and durability contrasted with the conventional (microscale) covering are a few instances of benefits of utilizing nanoscale covering in car motor.

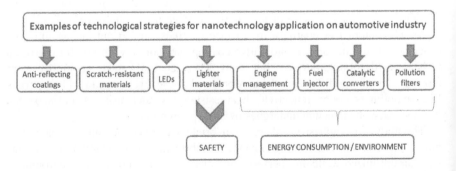

FIGURE 4.1 Instances for application of nanotechnology in the automotive industry (Coelho, M. C., Torrão, G., & Emami, N. 2012).

Car industry is continually putting forth attempts to the turn of events and application parts that are light weight, and which simultaneously have superb mechanical and tribological properties (Sharma, P., Khanduja, D., & Sharma, S. 2016). Today, the car business must meet certain necessities regarding lessening fuel utilization and simultaneously to keep up satisfactory solace and wellbeing of the vehicle (Suresh, S. et al 2014). Another significant factor is the heaviness of the vehicle, which directly affects fuel consumption and hence the discharge of toxic and harmful gases after combustion (Durowaye, S. I. et al 2017).

4.6.1 WEIGHT REDUCTION

The hardest aspect of the vehicle is the body casing, which makes 40% of its total weight. Subsequently, the most well-known way to deal with lightweight plan coordinated inside the structure of the vehicle which utilizes a blend of various materials, contingent upon their designing properties, and useful qualities. Vehicle mass is one of the most important design features that has a straight correlation with fuel consumption rate also. Decrease of 5% to10% in vehicle weight will prompt a decrease of CO_2 discharges between 1.3–1.8% and 2.7–3.6% individually.

Nanotechnology application into the car business prompts lighter vehicle bodies without bargain in firmness and crash obstruction implies less utilization of material and less fuel injecting because of decrease in weight. Progressed nanotechnology applications are permitting the executions of lightweight parts consequently bringing about vehicle decreased in weight. The normal 900 kg of steel and different metals in vehicles can be diminished by up to 300 kg using nanocomposites.

The essential materials used in the vehicle business are composites which are made with good quality steel, aluminum, and composites of carbon fiber and plastic. Applying top notch steel is cultivated by diminishing the weight by 20% appeared differently in relation to the conventional steel structures, use of aluminum of 40% with composites of carbon fiber. Henceforth, the use of light materials limits the weight of the vehicle (Jović, D., & Milićević, J. 2017).

4.6.2 EMISSION REDUCTION

In recent decades, the class of heterogeneous catalysts were utilized lessen outflows from vehicles motor burning. The catalysis response can be improved by utilizing chemically dynamic nanoparticles onto an exceptionally permeable help material with extremely high surface region for car emanation cleaning. A further use of permeable nanocomposites is connected with the utilization as contamination filters, which precisely or potentially by synergist response stifle discharge of ash particles or poisonous gases (Veličković, S. et al 2019). It has been discovered that expansion of Al or Al_2O_3 nanoparticles to diesel fuel improves its properties.

4.6.3 ENHANCEMENT OF ENERGY EFFICIENCY

Photovoltaic cells on the vehicle roof top are as of now a choice with certain impediments. To improve these cells execution, nanocomposites with semiconductor lattice can be utilized to improve adsorption and efficiency. Different nanocomposites showed up, for example, a photovoltaic paint made out of color-sharpened TiO_2

nanoparticles implanted in an electrolyte or a flexible slight film semiconductor cell with multinanolayers, or polymer cells either with carbon bucky balls or with semiconductor nano-rods. Based on an envisaged efficiency of 10%, one can pick up around 0.5 kW of electrical force for a limousine, enough to take care of battery. Even by deflecting the liners in car motor, the motor measurement diminishes significantly and decrease of payload is basically significant for fuel efficiency.

4.6.4 Nano-in-Tribology

Brake screech is notable issue brought about by unique precariousness in vehicles, which is influenced by vulnerability of its structure and little aggravation because of the variety in friction force. The brake viability, as a piece of brake execution, is required for vehicle wellbeing. The utilization of nanomaterials in planning new brake material has exhibited an expansion in brake's life time and decrease in brake screech. Indeed, even the wear and erosion in vehicle tires can be diminished by utilizing nanotechnology. Tire manufacturers are as of now adding nanoparticles like carbon black to improve wear to abrasion resistance and gas permeability (Smith, A. 2006).

4.6.5 Scratch-Resistant Paints

The utilization of hard, water-repulsing polymer nanocomposites or quartz nanoparticles empowers the coatings to stay clean themselves and secure against scratches and chips, and lessen corrosion, without modifying the presence of the paint underneath. Creating advancements incorporate coatings which are not simply more scratch-safe, even recuperate any minor harm which they do support. This is accomplished by implanting ceramic nanoparticles into polymer, which can stream over itself to mend breaks and scratches while staying hearty. Paints with nanoparticles can likewise be utilized to modify their warmth pondering properties depending on intensity of light incident. This assists with controlling the temperature of the vehicle, making the activity of the cooling framework simpler and, hence, saves fuel.

4.6.6 Adhesives

The adhesives have been utilized to upgrade welded joints and mechanical latches. These are improved with either iron oxide nanoparticles, or carbon nanotubes can supplant and outer form welds by and large. This encourages the metal to bond emphatically with plastic or composite boards, which diminishes weight and cost.

4.6.7 Fillers for Tires

Nanoparticles of carbon or silica have been utilized as fillers for tire elastic for a long time, assisting with upgrading tire execution and strength. Silica, which doesn't normally blend well in with elastic, is currently frequently utilized in concurrence with organosilanes, which can bind two parts.

4.6.8 Dirt-Resistant Paints for Windows and Wipers

The way to deal with dirt safe paints for vehicles, which utilizes nanoparticles to make a hydrophobic (water-safe) surface has additionally been applied to glass of vehicle windows. The beading impact for water on a superficial level implies that water runs off significantly more effectively and doesn't hinder perceivability during substantial downpour or splash, which, thus, lessens wear on windscreen wipers (Soutter, W. 2012). Nano-coatings on inside surface of glass can utilize a comparative way to deal with prevention of water vapor from condensation on the glass in moist conditions.

4.6.9 Ceramic Matrix Nanocomposites in Automotive

Ceramic composites with scattered metal particles speak to a class of promising material for elite applications in unforgiving conditions, for example, high temperature. These materials offer the chance of combining a resistance to heat, resistance to degradation, and the abrasion resistance of the ceramic phase having a high mechanical quality and thermal conductivity is furnished with metal stage.

Ceramic matrix nanocomposites are utilized for making nozzle assemblies, materials ovens, frameworks for changing over vitality, gas turbines, warm motors, and so on (Durowaye, S. I. et al 2017). These materials are unaffected with increase in temperature against steel, which changes its mechanical properties with increase in temperature. Specifically, they demonstrated great composite materials strengthened with carbon strands, which are profoundly sturdy at high temperatures, have a high wear opposition, and have considerably lower explicit weight contrasted and steel and aluminum. Carbon fiber strengthened composites are progressively being utilized for vehicle parts that are presented to high loads, for example, brake plates, valves, chamber liners, flash attachments, sensors, isolators, channels, cylinder, and others. Safe driving is accomplished by applying the brake discs made of ceramic composites, in light of the fact that the plates have an incredible protection from twisting and wear, in this manner empowering an extraordinary savings in vehicle weight.

4.6.10 Other Prospective Applications of Nanocomposites

Other potential applications of nanocomposites in the car business (Borgonovo, C. 2010) incorporate the top of belt gear unit, the cap, fuel hoses, fuel valves, entryway outlines, backrests, fuel hoses, and fuel valves, entryway outlines, seat backs, step helps, substantial electrical walled in areas, sail board, box rails, sash, barbecues, hood louvers, instrument boards, side trims, body boards, and bumpers, sensors, power devices. Figure 4.2 shows the use of polymer nanocomposites on different parts of car (Patel, V., & Mahajan, Y. 2012).

4.7 IMPACT ON REQUIREMENTS

Here we examine the effect of the necessities in automotive designing.

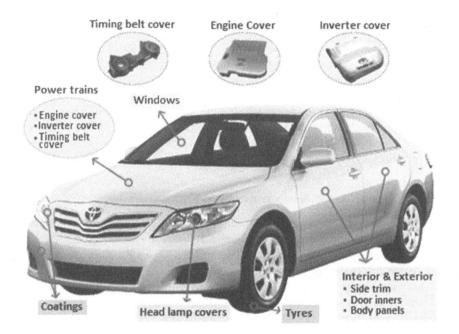

FIGURE 4.2   Use of nanocomposites in different parts of car (Patel V., & Mahajan, Y. 2012).

4.7.1   Electric Drive

From a car prerequisite designing point of view, electric drive is just about an "ordinary" mechanical development, similar to substitution of conventional instrument group needles by high-goal shows. Notwithstanding, on the idea prerequisites we saw that distinguishing the market's needs was (and is) testing. Which execution prerequisites (like range, charging time, greatest speed) at which cost is requested and acknowledged by the market? At the point when we have these limitations, framework and segment level necessities building is more "customary." For new kinds of car segments, similar to DC-DC converter, electric engine, or HV-battery, we needed to create sufficient quality prerequisites. Here we saw that before all else we would, in general, be excessively severe. As we are currently determining the fourth era of electric drive parts, we see a "standardization" of value necessities.

4.7.2   Network and Shared Services

Necessities designing difficulties in the network and shared and benefits field are multifaceted:

- For intra-vehicle needs, similar to basic association of buyer gadgets in the vehicle foundation, it is once more "standard." From an innovative point, the assortment of gadgets alongside the high change rate is obviously a test. In any case, this doesn't need new prerequisites building draws near. Plainly, necessities change the executives and prerequisites variation taking care of must be accomplished all the more carefully, yet there is nothing truly new.

- The online access of a driver to their vehicle (for example, to see the current charging status of the HV-battery or the vehicle's area) acquires prerequisites building innovations from the IT field. In this way, some report from a car RE point of view, however not from prerequisites designing all in all.
- Communication between vehicles (either legitimately or by methods for spine workers) suggests from a necessity building point of view primarily managing vulnerabilities (how dependable are the moved data?), security issues (is there an assault?), and normalization (the more accomplices are offering data, the higher is the advantage for all).
- Offering better approaches for getting brief admittance to portability, for example, by free gliding vehicle sharing or application upheld leasing of private claimed vehicles.

Fundamentally, we can reason that availability just as shared and administrations bring new prerequisites designing difficulties into the car business. Yet, from a RE point of view, there is nothing truly new.

4.7.3 Self-Sufficient

From a prerequisite designing viewpoint, with self-sufficient driving we are in reality entering another field. Here, a framework needs to collaborate with this present reality without a human fallback choice and no quick available safe state. The more we are pushing forward in that field, the more difficulties we face, as:

- Detecting the condition of a traffic signal can be a truly difficult issue if there are multilight (for example, for isolated paths or various headings).
- Overall, we face an obstruction object on motorways each 8,000 km. The response to such an article ought to be sufficient.
- Interaction with human street clients: At convergences, crosswalk, bottlenecks, to give some examples, the arrangement with others is frequently done by motion or other casual methods for correspondence.
- In basic circumstances (for example, a landslide on the motorway) the driver is mentioned by the police to carry on against clear guidelines (for example, utilize a disallowed motorway exit or drive back in opposition to the driving course.

From these models, we can undoubtedly observe that there are uncountable circumstances that can't be predicted and it is, hence, difficult to determine the ideal conduct definitely. Along these lines, great arrangement situated prerequisites designing methodologies are destined to come up shortly.

Plainly, we need to begin with objective-based prerequisites. However, here we regularly end up with logical inconsistencies, similar to the accompanying straightforward model shows:

- Goal 1: Do not pass a red traffic signal.
- Goal 2: If a crisis vehicle (for example, crisis specialist) approaches, empower its.

So bypassing we may run in the circumstance that we are holding up in front of a red traffic signal, when a crisis vehicle comes nearer from the back. Will we pass the red traffic signal? Will we move the vehicle on the asphalt? Will we hold up as the crisis vehicle has different choices? Also, if you don't mind note: We as of now have recognized this circumstance; nonetheless, there are numerous choices that must be assessed in the solid setting within reach.

There are significantly more circumstances that no one has just recognized. From a prerequisite designing viewpoint, we are confronting a totally new class of issues where no legitimate necessities building arrangements are accessible yet.

4.8 GENERAL ISSUES

4.8.1 PRINTED NECESSITIES ARE JUST ONE ASPECT OF THE GAME, CAR IMPROVEMENT IS TOO INTRICATE TO EVEN CONSIDER BEING OVERSEEN JUST BY TEXT

This doesn't mean, obviously, that literary prerequisites are not significant. Notwithstanding, handling and controlling only the literary prerequisites are plainly insufficient. Even all the more along these lines, with regards to complex frameworks, "controlling only the printed necessities" turns out to be almost outlandish without extra properties like working state (for example, starting, characterized, settled upon, delivered) or due date (for example, B-test, ..., etc.) reflecting qualities and periods of detail or advancement measure, separately (Robertson, S., & Robertson, J. 2012).

The significance of these and numerous different characteristics is currently all around recognized and helpful assortments exists (Wiegers, K. 1999). The precarious point is to tailor the correct arrangement of characteristics, to such an extent that the endeavors of characterizing and keeping up them influences with the advantages of better cycle control and determination reuse (Sommerville, I., & Sawyer, P. 1997). Upkeep of credits alludes both to the occasion level (for example, changing qualities and so forth) yet additionally on the class level (venture wide control of characteristic definitions) (Kotonya, G., & Sommerville, I. 1998).

The historical backdrop of prerequisites must be taken care of as well. This doesn't just incorporate account the "nearby" history of control of a solitary necessity. Architects commonly need to find in an all-around organized way the "worldwide" history of controlling a characterized assortment of prerequisites comparative with some past pattern (Gause, D. C., & Weinberg, G. M. 1989).

Notwithstanding qualities and history, in determinations, we additionally need to manage illustrative photos of some random configuration, tables, proposals of different sorts, clarifications about, for example, the structure method of reasoning or plan choices taken, test data, boundaries and other interface data of different sorts, and going with foundation data (passages from principles and so on). This sort of extra data frequently makes up undeniably over half of the general detail. It ought to be noticed that the greater part of these articles are not framework necessities themselves, as in they determine a necessary framework property. Be that as it may, they all need to have traits and history.

All these various articles will have different conditions (where here a "reliance" among A and B here implies that a difference in A may likewise prompt a difference in B). A portion of these conditions will stay certain; others will be made express in different manners in order to be efficiently recognizable (by fitting naming or by unequivocal connecting). We give a few encounters about express detectability beneath. The unpredictability of objects of various kinds, their qualities, and connections between them rapidly gets mistaking for the architect. Hence, there is a dire requirement for a metadata model that is characterized or adjusted previously and which officially characterizes this data. We discovered such metadata models to be of focal significance in necessities the executives' ventures: such a model speaks to the premise to examine with engineers both what their needs are and how they ought to methodically deal with their particulars. At DaimlerChrysler, we have presented the abbreviation RMI (Requirements Management Information model) for these models. The improvement of an RMI along with engineers has become the focal methods for a fruitful presentation of prerequisites the executives into an undertaking (John, G. et al 1999). For clear reasons of consistency and reusability, we have presented a vast secluded RMI, so new tasks can begin prerequisites the board exercises by adjusting and fitting this model. On the device uphold side, devices ought to be founded on an information base framework that understands the RMI. Furthermore, they ought to have a graphical proof reader for the RMI.

4.8.2 THERE IS NO REASONABLE LIMIT BETWEEN MAKER NECESSITIES DETAILS (LASTENHEFT) AND PROVIDER FRAMEWORK IN PARTICULARS (PFLICHTENHEFT)

In German, there is an exceptionally obvious qualification somewhere in the range of Lastenheft and Pflichtenheft. At first look, the English partner is the pair client prerequisite particular and framework necessity detail. The client necessities archive should determine what the issue is, while the framework prerequisites record ought to depict the answer to tackle the issue. As a rule, the client determination record determines what the framework to be created ought to have the option to do. The framework detail record characterizes how the engineer intends to understand the client prerequisites.

At second look, the connection between Lastenheft/Pflichtenheft on one hand and client prerequisites archive/framework necessities record isn't so natural. The Lastenheft is a detail record given by the client and mentions to the potential provider what the orderer needs to get. The Pflichtenheft as the provider's response to the clients list of things to get reports how the provider intends to acknowledge what the client needs to get. The issue is that it bodes well not to confine the

Lastenheft to the difficult space: Especially in our area, there are striking reasons why depicting the issue space just isn't sufficient:

- Systems are commonly partitioned into subframeworks and segments given by various providers, both inside one vehicle arrangement just as in successive ones. DaimlerChrysler needs to ensure that parts given by a few providers are going to cooperate both inside one vehicle arrangement yet in addition conceivably in comparable designs of future models. This implies, so as to indicate provider explicit "issues" we additionally need to determine portions of a general arrangement.
- When attempting to isolate programming from equipment parts or even programming providers from equipment provider's things get considerably harder: Now in any event, for a solitary segment, we need to determine two intelligent "issues," in particular, an equipment and a product one. Besides, the product framework ought to be as free from explicit equipment parts as could be expected under the circumstances.
- In numerous circumstances, DaimlerChrysler's first-to-advertise system along with the interest to keep serious skill in house requires for particular records that characterize certain highlights as itemized as a solid calculation can get, now and again gave as black box. Regularly, DaimlerChrysler should set the future norm by determining arrangements rather than issues. Some business relations between DaimlerChrysler and comparing providers are exceptionally old and have developed after some time. This occasionally prompts an exceptionally casual method of participation, in the most pessimistic scenario to finish reliance from the provider.

As a result, we some of the time find significant data missing in the Lastenheft. Or on the other hand, there are prerequisites recorded in the Lastenheft that ought not to be there, as they limit arrangement space in an expensive, yet exaggerated way. By and by, there are circumstances where even the prerequisites building master along with the space engineer has issues to choose whether a given necessity ought to be important for the Lastenheft or not. Through and through this prompts a circumstance where it is difficult to make a reasonable qualification between what ought to be important for the Lastenheft, the DaimlerChrysler engineer should convey and what ought not to be essential for this archive the limit somewhere in the range of Lastenheft and Pflichtenheft is a long way from self-evident.

4.8.3   EXCESS IS THE PLACE ACCEPTABLE SPECIALISTS ARE TRULY ENDURING—AND THE SUBSEQUENT FRAMEWORKS ARE PROBABLY GOING TO BE ENDURING, TOO

In huge activities, for example, the improvement of a car electrical framework, a few hundred enormous determinations and related archives are explained in equal under a difficult stretch timetable. It isn't amazing that ordinarily these reports contain a lot of excess data.

The impediments are self-evident: repetition prompts irregularities and twofold work. Or on the other hand—in the most pessimistic scenario—to detail reports got from invalid premises, prompting parts that don't satisfy their prerequisites or don't fit into the general framework.

The conditions between archives are perplexing and remain mostly in straightforward; this is just both because of the measure of material and faculty included and because of the absence of time.

It is extremely hard to manage this issue within the sight of report structures having developed and having picked up acknowledgment over numerous years. In addition, the record structure normally reflects hierarchical structures and even the whole car business structure. Changes in the record structure to eliminate redundancies are accordingly hard to accomplish. The compromise among endeavors and results is regularly negative here.

From our experience, it is now a significant advance to acquire and convey a review of the redundancies inside the undertakings. Designers, hence, become mindful about those pieces of their report being utilized or expounded in equal some spot else and they can choose to synchronize if important.

A definitive extremely long haul arrangement, obviously, is an effective information base of detail related articles, disseminated over makers and providers, from which all records are produced.

4.8.4 Content Isn't the Place Acceptable Designers Are Enduring—Introduction of Substance and Nearby Legends, Notwithstanding, May Prompt Issues

In the car space—however, presumably in most inserted framework areas—frameworks are normally not based on the green field yet in increases: the new vehicle arrangement acquires most usefulness from previously existing ones with pretty much variations, augmentations, or developments. The new windshield wiper fundamentally is simply one more (and surprisingly better) windshield wiper.

As an outcome, at this level, there is no proper elicitation and arrangement stage, just moderately not many related exercises are important concerning the little augmentation being created in another variant of a part. We found that, from numerous long periods of experience, great architects are head class specialists in their specific space and have acquired profound understanding on the nuances and traps of the necessities content. For them, substance and space information isn't the primary issue.

Be that as it may, there can be obviously issues identified with the introduction and organizing of substance. Frequently, engineers don't portray details in a well-structured and effectively open way, prompting high correspondence endeavors inside a producer or between a maker and a provider. We contend here that a very much characterized RMI actualized by a prerequisites the board instrument can assist with improving the organizing and introduction of substance, for instance, by pushing designers to separate (extensive) detail sections into nuclear necessities with related tests.

There is another purpose of progress: the reuse of old particulars today happens in a specially appointed and understood manner, which is neither productive nor entirely attractive as for detail quality. The primary issue determination is totally subject to the one specialist being the master in the given space for a considerable length of time (Cybulski, J. L., & Reed, K. 2000). It tends to be helpful to respect precise reusing of existing particulars as an unequivocal advance of the detail cycle (Wiegers, K. 2001). From a necessity designing cycle perspective, we will return to this point in talking about the following articulation underneath (Lam, W., McDermid, J. A., & Vickers, A. J. 1997).

A firmly related specialized issue while presenting a prerequisite the board information base is to relocate existing reports to make them more reusable for up-and-coming undertakings. While prerequisites the executive's apparatuses offer some programmed catching usefulness, we found that a ton of manual work is as yet important to relocate archive content into a necessity the board information base so it turns out to be more reusable. Moreover, various undertakings must utilize viable prerequisites the executive's data models so as to accomplish cross-venture reusability. This again focuses on the requirement for a cross-venture normalized RMI when creating item families.

4.8.5 SHAPED PLATES

Formed plates expect significance in the plan of protected individual automobiles. These plates are used for impact alleviation erections against land mine impacts. The molded plates redirect and retain specific aspect of the verve granted to them. Impact relieving limit at a specific drive level basically relies upon material and the mathematical boundaries of the plate viable. The significant boundaries of intrigue are the mass of the touchy and the included point of the V molded plate. Short proximity shoot loads from land mines represent a serious danger to the military vehicles and common framework. In the majority of the cases, the heaps are hard to foresee since countless autonomous factors are found to influence the stacking. Joining molded metal plates underneath the vehicle floor is the main methods for relieving the impact of the shoot loads in vehicles. In military automobiles, giving a V shape structure is the most normally utilized technique for alleviating the impact impacts. When contrasted with a level plate, the V molded plate avoids an aspect of the impact load and assimilates the rest of the part to guarantee adequate degree of security. A wide range of states of the plate like level shape, V shape, allegorical shape, and so on are accessible yet V shape gives the ideal security.

4.8.6 SOLUTIONS TO THERMAL ISSUES IN AUTOMOTIVE ENGINEERING

A wide assortment of materials, gadgets, and frameworks have been developed and improved in the course of the most recent couple of years to help upgrade vehicle warm administration systems in every one of the three territories generally essential to the field: the lodge, power hardware, and outside. The execution of develop advancements and further innovative work into forthcoming arrangements vows to yield noteworthy advantages—not exclusively to makers and purchasers, however to society all in all.

The fundamental difficulties in the vehicle lodge are identified with the HVAC framework, which puts a weighty interest on the vehicle main player (e.g., the motor or battery), in this manner decreasing efficiency and range. Enhancements for the most part center around two territories: diminishing warmth ingestion to bring down the HVAC load and improving the productivity of the cooling framework. Advancements, for example, surface coating and coloring have been appeared to lessen the warmth load on a vehicle when it is presented to the sun. Nonetheless, colored glass is changeless, and when the climate is chilly, warm douse is alluring to warm the lodge utilizing free sun oriented force. Photograph Volta chromic gadgets ought to be concentrated further to permit warm splash to be fluctuated relying upon encompassing conditions. Zoned cooling advances can diminish power requests on HVAC frameworks. Clients have as of late discovered algorithmic improvement of ACC frameworks to be more helpful than frameworks of the past. Central air frameworks in EVs must be managed uniquely in contrast to those in regular ICEs.

Execution of warmth siphons and warm batteries was examined as a way to decrease the force request from the battery pack, and further examination in these zones is justified. Warmth siphons keep on being generally productive in colder atmospheres, yet have just been inspected for the extraordinary instance of warming electric transports, wherein the warmth yield is more noteworthy than in littler vehicles. More examination into the exhibition of these gadgets in EVs, just as the attainability of working such gadgets in cooling modes in more sultry atmospheres, might be justified. Warm batteries stay in their early stages, so contemplates including their effective incorporation into EVs ought to be sought after and endeavors to improve their life span ought to be made. A few nations, for example, the United States, have a wide scope of atmospheres. Along these lines, vehicle producers should zero in on creating lodge warm administration frameworks that are suitable for a scope of conditions. Future investigations should remark on the framework's versatility; if a framework is just appropriate for a specific district, it should exist as an "attachment and-play" segment that can be handily consolidated into a huge determination of cars. As a rule, lodge warm administration considers will, in general, spotlight on either warming or cooling the inside of a vehicle. Examination into how to best execute both into a solitary vehicle will at last be important. Warm splash can be tended by means of ventilation, utilizing techniques, for example, floor vents. Be that as it may, exhaust items, soil, and creatures may enter the lodge through these vents. Subsequently, new techniques should be investigated that give ventilation to improve lodge temperature while maintaining a strategic distance from unwanted interruptions and different issues. Ventilation fans have been found to lessen warm burden successfully, yet can have disadvantages like those of floor vents. They likewise have potential productivity issues, which can be tended to using sunlight-based force.

On overcast days, less force would be accessible, yet without direct daylight, warm entrance ought to likewise be lower. Traveler solace may likewise be expanded by discovering approaches to control inside dampness by means of ventilation frameworks. As hardware in vehicles become littler, all the more remarkable, and more significant—particularly in EVs—better cooling procedures become basic. The two significant segments of concern are the battery and the IGBT. Inactive cooling of the IGBT utilizing swaying heat pipes has been researched, and the outcomes exhibit that $(CH_3)_2CO$ and n-pentane are reasonable working liquids for this application. Thermal Interface Materials can more productively move heat from electronic parts to inactive coolers. Dynamic cooling by means of AHSs, which move heat through a fixed small circle of fluid metal compound, has been exhibited to be a doable hardware cooling arrangement. Another sort of dynamic cooling is fly impingement, in which heat-disseminating surfaces come in direct contact with the coolant for higher warmth move coefficients.

Strategies to cool the battery pack incorporate constrained convection, heat funnels, and PCMs. Warmth pipes are best when incorporated with the battery divider and can likewise be intended for either free or constrained convection cooling. Notwithstanding, heat pipes have not yet been actualized in a completely working framework for an all-inclusive timeframe. Examination on improving warmth pipe unwavering quality may give a more steady, lightweight, and inactive arrangement. Extra examination on the fast assembling and cost decrease of warmth channels would likewise demonstrate gainful for large scale manufacturing. Improving EVs and HEVs will require finding better approaches to disperse warm vitality from the battery pack and IGBT, just as tending to warm structure constraints. The vehicle outside contains numerous segments and warmth sources that influence the remainder of the vehicle. The surface temperature is firmly affected by direct daylight, while wind current through the flame broil influences radiator execution and reduces the mileage. Fan sizes and shade frameworks have been mimicked and tried to tackle the last issue. Endeavors could likewise be made to lessen haul by discovering ideal methods of coordinating wind current in the engine and tire well areas. Vehicle brakes create high warmth motions during the slowing down cycle. Wheel talked tendency and ventilated rotors have been found to eliminate and move the warmth most successfully. Tire distortion influences tire temperature, and at high speeds, the disfigurements become more serious and cause high temperatures and weight on the tires. Numerous re-enactments have been done to contemplate hysteresis, yet the field would profit by utilizing this data to overhaul tires that are, G. J. Marshall et al./Engineering 5 (2019) 954–969 967, more successful. The underbody of the vehicle is influenced by heat from the fumes framework. Recreations have demonstrated the impacts of fumes heat, yet these reenactments can be made more precise by incorporating transient temperatures related to the hot gas coursing through the fumes, as opposed to streamlining the fumes pipe as a steady temperature heat source. Regenerative slowing down is another innovation in which full-scale warm examination is required. Brake recreations should likewise consider the warmth move in both the brake cushion and the rotor. Reproductions that fuse the cyclic idea of the inner burning motor as a warmth source would be of extraordinary intrigue.

4.9 APPLICATION ZONE

With expanding requests on all capacities in the vehicle, the control and guideline works that were recently actualized by mechanical methods have, over the most recent 40 years, progressively been supplanted by electronic gadgets (ECU). This has essentially caused appeal for sensors and actuators with which these electronic control units can, from one viewpoint, decide the pertinent conditions of the vehicle, and then again, really impact these states.

Over this period, the vehicle business has gotten one of the, beforehand extraordinary, drivers of the advancement of sensors that could be fabricated in enormous numbers. While at the start, they despite everything were of electromechanical or large-scale mechanical type or some likeness thereof, the pattern starting during the

80s was for scaling down, with semiconductor techniques (group preparing) for the production of high use sensors.

Sometimes sensors beginning in half and half innovation utilizing thick-film procedures likewise assumed a not deficient job. These are as yet utilized in confined applications today, for example, in the wafer-molded Lambda oxygen sensors and high temperature sensors for the fumes line. Where temperature and attractive field sensors could at first despite everything be planned in circuit-like structures and be delivered in bunches, this pattern expanded as it became conceivable likewise to micromachine silicon from various perspectives in a few measurements, and furthermore to interface them, even in numerous layers, adequately by exceptionally productive strategies.

The advances of the electronic semiconductor circuits were essentially only dependent on silicon as the base material, very various materials, and innovations assumed a not inconsequential function in sensors. Along these lines, for example, quartz can likewise be micromachined utilizing anisotropic drawing procedures, and not at all like silicon additionally has extremely invaluable piezoelectric properties. III-V semiconductors, for example, gallium arsenide (GaAs) have a generously more prominent working temperature compare to silicon, which can be extremely beneficial, especially at certain focuses in the vehicle. Slight metallic layers are very appropriate to the production of exact strain-gage resistors, precise temperature sensors, and attractive field-subordinate resistors.

With silicon it is conceivable additionally to incorporate the gadgets solidly with the sensor. This method has lost its significance—with a couple of special cases (for example, Hall Effect ICs)—on account of the by and large altogether different number and kind of cycle steps and furthermore as a result of the rigidity related to this. Crossover mix innovations in the most impenetrable of spaces, generally speaking, lead to considerably more practical, however practically equal arrangements.

While the improvement of sensors was at first focused solely on frameworks inside the detail in the drivetrain, the suspension, and the body and driving well-being, the bearing of detecting of more up to date advancements is progressively named to the outside and to the region near and further from the vehicle: Ultrasound sensors recognize obstructions on leaving and will even permit have programmed leaving within a reasonable time-frame, maybe in mix with different sensors.

- Near-run radar examines the territory around the vehicle to identify objects that most likely could influence a crash, to pick up time, and to prime security frameworks before an impact happens (precrash sensors).
- Imaging sensors cannot just recognize traffic signs and send them to the driver's presentation, yet in addition distinguish the edge of the carriageway, caution the driver of any risky deviation and, where required, in the long haul likewise license programmed driving. In blend with infrared shafts and a screen in the driver's field of vision, IR-touchy imaging sensors could allow significant distance perception of the carriageway, even around evening time (night vision) or in foggy conditions.
- Long-extend radar sensors watch the carriageway for 150 m before the vehicle, even in helpless perceivability, to adjust the driving pace to vehicles ahead and, in the more drawn-out term, additionally to help programmed driving.

As a feature of the vehicle's fringe hardware, the sensors and actuators structure the vehicle's interface to its unpredictable drive, slowing down, undercarriage, and bodywork capacities, just as to the vehicle direction and route capacities and the (normally computerized) ECUs which work as the handling units. A connector circuit is commonly used to change over the sensor's signs into the normalized structure (estimating chain, estimated information enrolment framework) required by the control unit.

These client explicit connector circuits are customized for explicit sensors and are accessible in incorporated plan and in a wide assortment of forms. They speak to a very significant and truly important expansion to the sensors portrayed here, without which their utilization would not be conceivable, and the estimating exactness of which is, appropriately stated, just characterized related to these.

The vehicle can be viewed as a profoundly intricate cycle, or control circle, which can be affected by the sensor data from other handling units (control units), just as from the driver utilizing his/her controls. Show units keep the driver educated about the status and the cycle overall.

4.10   SUMMARY

Over the past decade, technology has taken a huge leap and this has ensured to make man life easy. Technology has shown its effects in all areas and mainly in the field of automotive. Technology has ensured in making car driving experience safer and comfortable. This is possible by integrating electronics into a single system with better communication between the sensors in the car. In present day, the use of cloud has enabled the data from the sensor to be analyzed in a proper way in order to understand the behavior of the car in different environment. As the day's progress the advancement in automotive technology is an ideal collaboration between man and machine.

This technology has spread its roots from designing, manufacturing, and assembling the cars. Companies like google, tesla, IBM have all worked in tandem with each other directly or indirectly in making the automotive industry the place where it is today. In the future, the advent of electric drives looks into to green environment and clear air. All this is possible with technology spreading its wings in a rapid rate.

Nanomaterials have been utilized for a wide range of car applications, similar to wellbeing, decrease in vitality utilization, expanding quality, decrease weight, etc. It must be underscored that the vehicles' weight decline is a reasonable outcome between the use of nanomaterials and the utilization of structures produced using low-density materials. Indeed, even nanotechnology will consider the more efficient utilization of materials, just as vitality and accordingly will lessen waste and contamination. For instance, by utilizing a limited quantity of noble metals on transporter material of exhaust emanation impetuses, the discharges of hydrocarbons, carbon monoxides, and nitrogen oxide could be diminished by 90%.

The benefit of utilizing nanocomposites in outer parts and drive extravagance vehicles positively affect market demand. Nanocomposites dependent on an assortment of materials of metal or plastic strengthened with metal or ceramic nanoparticles can essentially improve the quality of parts. Various prerequisites of the car business, like improved mechanical, electrical, thermal, corrosion, self-cleaning and anti-wear properties, and sensing capacities strengthened with nanoparticles, nano-films, nanoflakes, nanotubes and nanofibers as nanocomposites. This area has also used all the combined modern technologies with human thirst for excellence, knowledge, and every step up the ladder of science and even the wars that provided us the tools have now been used in cars like antennas, radars, and wireless communication to ensure a safer driving experience. There by making is a self-dependent field.

REFERENCES

(Bajaj, P., & Khanapurkar, M. 2012) Bajaj, P., & Khanapurkar, M. 2012. Automotive networks based intra-vehicular communication applications. In New Advances in Vehicular Technology and Automotive Engineering (pp. 207–230).

(Borgonovo, C. 2010) Borgonovo, C. 2010. Aluminum nano-composites for elevated tempera-ture applications. Master of Science Thesis, Woreester Polytechnic Institute.

(Coelho, M. C., Torrão, G., & Emami, N. 2012) Coelho, M. C., Torrão, G., & Emami, N. 2012. Nanotechnology in automotive industry: research strategy and trends for the future—small objects, big impacts. *Journal of Nanoscience and Nanotechnology*, 12(8), 6621–6630.

(Cybulski, J. L., & Reed, K. 2000) Cybulski, J. L., & Reed, K. 2000. Requirements classifica-tion and reuse: crossing domain boundaries. In International Conference on Software Reuse (pp. 190–210). Springer, Berlin.

(Devi, R. S., Sivakumar, P., & Sukanya, M. 2018) Devi, R. S., Sivakumar, P., & Sukanya, M. (2018). Offline analysis of sensor can protocol logs without can/vector tool usage. *International Journal of Innovative Technology and Exploring Engineering*, 8(2S2) pp. 225–229.

(Durowaye, S. I., Sekunowo, O. I., & Lawal, A. I 2017) Durowaye, S. I., Sekunowo, O. I., Lawal, A. I., & Ojo, O. E. 2017. Development and characterisation of iron millscale par-ticle reinforced ceramic matrix composite. *Journal of Taibah University for Science*, 11(4), 634–644.

(Gause, D. C., & Weinberg, G. M. 1989) Gause, D. C., & Weinberg, G. M. 1989. Exploring requirements: Quality before design (Vol. 7). Dorset House, New York.

(John, G., Hoffmann, M., Weber, M. 1999) John, G., Hoffmann, M., Weber, M., Nagel, M., & Thomas, C. 1999. Using a common information model as a methodological basis for a tool-supported requirements management process. In INCOSE International Symposium (Vol. 9, No. 1, pp. 1437–1441).

(Jović, D., & Milićević, J. 2017) Jović, D., & Milićević, J. 2017. Influence of Application of New Material in Automotive Industry on Improving Quality of Life. In 2nd International conference on Quality of Life, Center For Quality, Faculty of Engineering, University of Kragujevac (pp. 331–332).

(Kotonya, G., & Sommerville, I. 1998) Kotonya, G., & Sommerville, I. 1998. Requirements engineering: Processes and techniques. John Wiley & Sons, Inc.

(B. L. Kovitz, 1998) Kovitz, B. L., 1998. Practical software requirements. Manning Publications.

(Lam, W., McDermid, J. A., & Vickers, A. J. 1997) Lam, W., McDermid, J. A., & Vickers, A. J. 1997. Ten steps towards systematic requirements reuse. *Requirements Engineering*, 2(2), 102–113.

(Lin, Y. C., Lee, W. J., & Chao, H. R 2008) Lin, Y. C., Lee, W. J., Chao, H. R., Wang, S. L., Tsou, T. C., Chang-Chien, G. P., & Tsai, P. J. 2008. Approach for energy saving and pol-lution reducing by fueling diesel engines with emulsified biosolution/biodiesel/diesel blends. *Environmental Science & Technology*, 42(10), 3849–3855.

(Patel, V., & Mahajan, Y. 2012) Patel, V., & Mahajan, Y. 2012. Polymer nanocomposites drive opportunities in the automotive sector. Nanowerk. https:llwww.nanowerk.com/ spotlight/sootid=23934.php. (Accessed by 10 June 2021).

(Presting, H., & König, U. 2003) Presting, H., & König, U. 2003. Future nanotechnology developments for automotive applications. *Materials Science and Engineering: C*, 23(6–8), 737–741.

(Robert Bosch Gmbh 2013) Robert Bosch Gmbh. 2013. Bosch automotive electrics and automotive electronics", Systems and components, networking and hybrid drive, 5ᵗʰ edition. Springer Vieweg.

(Robertson, S., & Robertson, J. 2012) Robertson, S., & Robertson, J. 2012. Mastering the requirements process: Getting requirements right. Addison-Wesley.

(Sandhya, D. R., Sivakumar, P., & Balaji, R. 2019) Sandhya, D. R., Sivakumar, P., & Balaji, R. 2019. AUTOSAR architecture based kernel development for automotive application. In International Conference on Intelligent Data Communication Technologies and Internet of Things (ICICI).

(Sharma, P., Khanduja, D., & Sharma, S. 2016) Sharma, P., Khanduja, D., & Sharma, S. 2016. Dry sliding wear investigation of Al6082/Gr metal matrix composites by response surface methodology. *Journal of Materials Research and Technology*, 5(1), 29–36.

(Sivakumar, P., Vinod, B., Devi, R. S. 2015) Sivakumar, P., Vinod, B., Devi, R. S., & Nithilavallee, S. P. 2015. Model-based design approach in automotive software and systems. *International Journal of Applied Engineering Research*, 10(11), 29857–29865.

(Smith, A. 2006) Smith, A. 2006. Does it have a sporting chance?. *Chemistry International*, 28, 8–9.

(Sommerville, I., & Sawyer, P. 1997) Sommerville, I., & Sawyer, P. 1997. Requirements Engineering: A good practice guide. John Wiley and Sons.

(Soutter W. 2012) Soutter W. 2012. Nanotechnology in the Automotive Industry, Azo Nano, https://www.azonano.com/article.aspx?ArticleID=3031 (Accessed by 10 June 2021).

(Subburaj, S. D. R., Kumar, V. V., Sivakumar, P. 2021) Subburaj, S. D. R., Kumar, V. V., Sivakumar, P., Kumar, B. V., Surendiran, B., & Lakshmi, A. N. 2021. Fog and edge computing for automotive applications. In Challenges and Solutions for Sustainable Smart City Development (pp. 1–15). Springer, Cham.

(Suresh, S., Moorthi, N. S. V., & Vettivel, S. 2014) Suresh, F. S., Moorthi, N. S. V., Vettivel, S. C., & Selvakumar, N. 2014. Mechanical behavior and wear prediction of stir cast Al–TiB2 composites using response surface methodology. *Materials & Design*, 59, 383–396.

(Swetha, S., & Sivakumar, P. 2021) Swetha, S., & Sivakumar, P. 2021. SSLA based traffic sign and lane detection for autonomous cars. In 2021 International Conference on Artificial Intelligence and Smart Systems (ICAIS) (pp. 766–771).

(Veličković, S., Stojanović, B., Ivanović, L. 2019) Veličković, S., Stojanović, B., Ivanović, L., Miladinović, S., & Milojević, S. (2019). Application of nanocomposites in the automotive industry. *International Journal of Mobility and Vehicle Mechanics*, 45(3), 51–64.

(Wiegers, K. 1999) Wiegers, K. 1999. Software requirements: A pragmatic approach. Microsoft Press, Redmond.

(Wiegers, K. 2001) Wiegers, K. 2001. Requirements when the field isn't green. *Software Testing and Quality Engineering*, 3(3) pp. 2–10.

5 Software Architecture for Autonomous Trouble Code Diagnostics in Smart Vehicles

R. Rajaguru
Department of CSE, Sethu Institute of Technology,
Virudhunagar, India

M. Mathankumar and T. Viswanathan
Department of EEE, Kumaraguru College of Technology,
Coimbatore, India

M. Manimaran
Department of EEE, Sethu Institute of Technology,
Virudhunagar, India

CONTENTS

5.1 INTRODUCTION

The issue in today's cutting-edge vehicles is that they have a wide range of trend-setting innovations yet come up short on the most significant thing, i.e., today's advanced vehicles can't self-analyze and report the issues to the client. A specialist

DOI: 10.1201/9781003269908-5

is needed to analyze the issue. Even though a few vehicles are fit for the conclusion. On-board analytic frameworks assume a critical job in the presentation of vehicles. For the most part, it takes more effort to analyze an issue in a vehicle than to correct it. So, these frameworks analyze the shortcomings as well as spare a ton of time. Boundaries watched and analyzed by the on-board diagnostic (OBD) framework are utilized by the electronic control unit (ECU) of the vehicle, which, thus, controls the motor's primary activity (like flash planning control, air-fuel blend, fuel injectors splashing period, and so forth.). A blend of these two frameworks guarantees security and proficient vehicle activity. Vehicles have an unpredictable structure comprising of subsystems, for instance, gearbox, motor, and brakes. The sensors and actuators are related and controlled with an ECU. It is related to controller area network (CAN), which has the unmistakable subsystems (Devi, R. S., Sivakumar, P., & Sukanya, M. 2018.). A significant level symptomatic convention is expected to speak with ECU. Two notable conventions which have the standard are OBD2 and UDS. OBD2 implies a vehicle's self-demonstrative and detailing capacity. OBD structure gives the vehicle proprietor to comprehend the current information of the vehicle and to get to the issue with away from and proficient admittance to the state of existing information of various vehicle subsystems while UDS gives all subtleties.

Since 1999, all vehicles must follow the international vehicle principles, i.e., all the vehicles fabricated from 2002 must have an OBD framework. The current framework doesn't have an appropriate gadget to cooperate with the demonstrative arrangement of the vehicle. However, it tends to be changed with the assistance of a driver's principle PC interfaced with the vehicle OBD. On-board analytic frameworks assume a basic job in the current age of vehicles and will assume an undeniably significant job in the following future. Sensors are utilized in regular items, for example, contact touchy, lift catches (material sensors), and lights, which diminish (or) light up. It is utilized to identify and react to some sort of contribution to the physical condition. The particular data could be light, heat movement, weight, and dampness. This essential capacity is utilized to detect the vehicle's feeling (or) some other wrong conditions.

A keen vehicle framework can be utilized for the current and cutting-edge vehicles. to accomplish improved riding experience and recognizing the issues. Shrewd vehicle PC frameworks can be isolated into driver focal PC frameworks and symptomatic frameworks. The driver's essential PC framework can go about as the primary wellspring of data for the driver about vehicle conditions and the general condition. The information which is recovered from OBD and sensors in the vehicle will be sent to the driver's primary PC and will be appeared on the heads-up show.

5.2 BACKGROUND

The present vehicles have increasingly more gadgets, and this prompts a gigantic abundance of data accessible on a vehicle's correspondence to arrange. Regardless of whether you are the improvement engineer, the business technician, or the end-client, the data accessible to the various people has gotten basic. Quick diagnostics of occasions and blames programmed restorative collaboration of vehicle hardware to remunerate disappointments, just as preventive support cautions of explicit vehicle segments are on the whole activities that are helping the different people all through the vehicle's presence.

Present-day vehicles having the framework as an underlying model for certain vehicles. This gadget goes about as a driver's principle PC and gives a factual and constant examination of the vehicle. There are a few components that are causing mishaps out and about. Those variables are human blunder, vehicle mistake, and the mishap itself, which can't be dodged (Ruddle, A. R. et al. 2013). The wellspring of mishaps in light of vehicle blunder is brake glitch, old vehicle, and the motor, which isn't looked after accurately. The ELM327 is an understanding chip which depends on RS232 Interpreter. Utilizing this interface, a few vehicle issues can be recognized. It may be utilized to get analytic data from the vehicle. This framework gives ongoing vehicle sensors. It is a minimized structure that can be fitted into any model vehicles (DiGiampaolo, E., & Martinelli, F. 2011). It might require power flexibly in any event, when the vehicle is killed. A few vehicles don't uphold the continuous determination

This model can be additionally improved with the goal that we can see the model from a versatile gadget. It may be additionally moved up, to make it ready to communicate with the on-board symptomatic system of the vehicle. This gadget can be interfaced with the CAN transport for more control in current vehicles. On-board symptomatic frameworks assume a fundamental job in the current age of vehicles and will assume an undeniably significant job in the following future. On-board symptomatic frameworks assume a vital job in the presentation of vehicles. Normally, it takes more effort to analyze an issue in a vehicle than to correct it. So, these frameworks analyze the issues as well as spare us a ton of time. Boundaries watched and analyzed by the OBD framework are utilized by the ECU of the vehicle, which, thus, controls the motor's primary activity (like flash planning control, air-fuel blend, fuel injectors showering period, etc.). A mix of these two frameworks guarantees wellbeing and proficient vehicle activity.

The OBD standard determines the sort of analytic connector and its pinout, the electrical flagging conventions accessible, and the informing design (Gallanti, M. et al. 1989). It additionally gives a competitor rundown of vehicle boundaries to screen alongside how to encode the information for each. Checked things incorporate speed, rpm, start voltage, and coolant temperature. This framework likewise advises a designer when an individual chamber has a fizzle. The OBD-II standard additionally gives an extensible rundown of diagnostic trouble codes (DTCs) (Hu, J. et al. 2010). Because of this normalization, a solitary gadget can question the on-board computer(s) in any vehicle.

OBD frameworks give the vehicle proprietor or a fix specialist admittance to the condition of wellbeing data for different vehicle subsystems. The Society of Automotive Engineers (SAE) built up a lot of principles and practices that controlled the improvement of these symptomatic frameworks. The SAE developed that set to make the OBD-II guidelines. The EPA and the California Air Resources Board (CARB) embraced these guidelines in 1996 and ordered their establishment in all light-obligation vehicles (Hu, J. et al. 2010). An outer programming unit and equipment particular are needed to get to the OBD framework. Additionally, the data given by the OBD is still in the hex-code design. At the point when the OBD framework identifies a breakdown, OBD guidelines require the ECU of the vehicle to spare a normalized DTC about the data of brokenness in the memory. An OBD scan tool for the servicemen can get to the DTC from the ECU rapidly and precisely to affirm the breaking down qualities and area following the prompts of DTC that abbreviates the administration time generally. Additionally, as of now, the quantity of thing for the ongoing driving status that OBD can screen is as high as up to 80 things

The Android gadget running with API 16+ (Jellybean to any most recent adaptation) can be interfaced with a pi running raspberries or any Linux working framework (Srihari, M. M., & Sivakumar, P. 2018). This can be an immediate correspondence between these two gadgets, which empowers information and data sharing through any sort of wired or remote network (Shiwu, L. et al. 2011). One of the fundamental and secure network conventions is the SSH customer (Secure Shell). The SSH convention is a strategy for secure distant login starting with one PC then onto the next. It gives a few elective choices to solid verification, and it ensures the interchanges security and uprightness with solid encryption. It is a safe option for the noninsured login conventions and uncertain record move techniques.

Intelligent and robotized document moves, giving far off orders, overseeing system foundation, and other strategic framework parts. Utilizing the SSH application for android, we can interface with any close by gadgets with a wired or remote system. It tends to be utilized to send shell orders to the associated gadget. The associated gadget can be legitimately associated and distantly controlled. It is one of the most secure conventions which incorporate confirmation (Babu, M. S., Anirudh, T. A., & Jayashree, P. 2016). Two techniques should be in a similar system to have the option to associate and speak with one another. Distant association requires a mind-boggling system arrangement. There is no appropriate graphical user interface that is accessible for this technique. The android gadget requires a customer for better correspondence. The correspondence convention can be improved in a manner with the end goal that it empowers the client-worker correspondence model. Rather than wired or remote a similar system association, we can utilize a worker for easier far-off availability. Along these lines, two gadgets can be implication associated (Sun, Q. 2002) (Stefanowski, J. 2013) (Tunggal, T. P., Latif, A., & Iswanto 2016).

Current-voltage converter, voltage–controlled oscillator, switch control circuit, and microcontrollers are the regularly utilized segments by utilizing a reference signal, the steady part can be estimated (incorporate counterbalance voltage). Exactness, dependability, low-cost, and effectiveness are demonstrated (Lu, W., Han, L. D., & Cherry, C. R. 2013) (Zhang, Y. et al. 2009).

5.3   SYSTEM ARCHITECTURE AND COMPONENTS

The overall architecture of the proposed system is shown in Figure 5.1. The proposed system is designed using the Internet of Things (IoT) and artificial intelligence (AI) techniques to monitor and control all the parts in the car. Our proposed work is

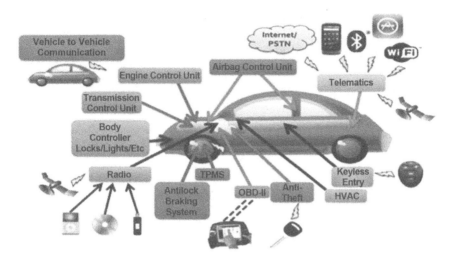

FIGURE 5.1 Car health monitoring architecture.

mainly concentrating on following the air-conditioning system and assists the user in identifying the problem that occurs in the car without mechanical assistance.

The proposed system's primary task is to maintain the vehicle's health report in regular time intervals to assists the users of the vehicle and monitoring all the electronics components and to operate the air-conditioning system over the internet. Our network consists of an OBD, driver's main computer, diagnostic sensors, a cloud server, an object detection camera, and an application developed in the android platform. The communication between all the components in that vehicle uses a near-field communication aspect over the internet. The driver's main computer will act as a central node that can communicate to the sensors and other electronic parts present in the vehicle as well as collects all the information from all the sensors and electronic devices (Shiwu, L. et al. 2011). The collected data will store in a cloud server for further processing. For monitoring and controlling all the sensors and electronic devices, an interface is needed, and that interface communicates to an ECU in that vehicle. The system uses a microcontroller, which acts as an interfacing unit in our system; we use the Pi microcontroller for interfacing purposes (Sandhya, D. R., Sivakumar, P., & Balaji, R. 2019). Figure 5.2 shows the connection between OBD, sensors, actuators, and the ECU. This connection is focusing on displaying the troubleshooting codes to the user, which is not in the existing system. In the existing system, if the problem occurs, then the unit will indicate the trouble by lighting the LED only, which is not assisting the user.

Figure 5.3 shows the system design, which focuses on monitoring and controlling the usages of the car and to identify the troubles that occurred in the car using the OBD codes such as P0300, P0654, P0700, P0049, P0039, P0900, and P0470. In this approach, all the electronics and sensors such as coolant temperature, pressure, air intake, fuel level, RPM, and ignition voltage sensors were connected to OBD as well as main driver computers because of the status and running conditions of all the electronics components of the vehicle communicate through the users.

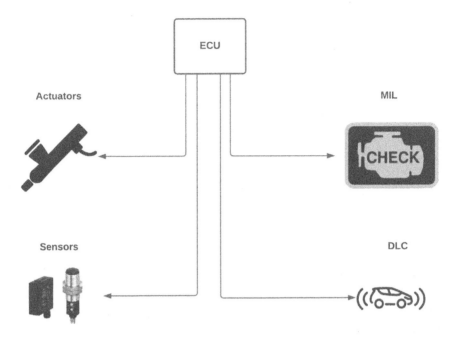

FIGURE 5.2   Connections with ECU.

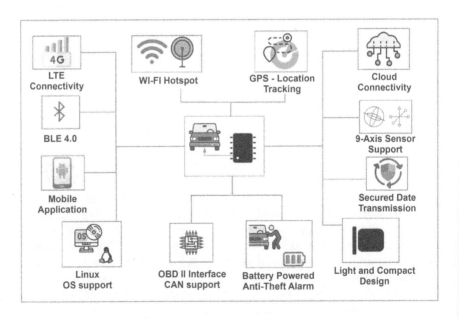

FIGURE 5.3   System design to monitor health record.

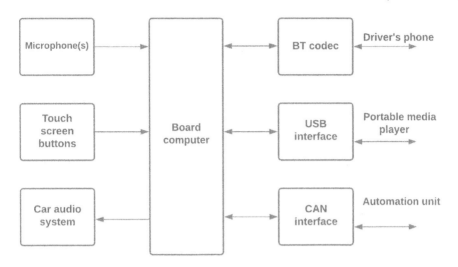

FIGURE 5.4   Driver's main computer.

To display the pieces of information as well as to indicate the troubleshooting codes, the electronic devices and sensors will be communicating their status to the user in a regular time interval as well as the user's need. To make this communication efficient, the proposed system uses the device-to-device communication using unlicensed channels in the industrial, scientific, and medical (ISM) band. Through this device to device (D2D) communication channels, it updates all the information to the user. The system gets data from different sensors fitted onto the vehicle and observes those parameters. The system will not only examine the parameters of the vehicle continuously but also display real-time values, warnings, etc., to the user. So, while driving, the user can easily monitor the happening in selected parts of the vehicle.

The driver's main computer acts as an interfacing unit between the electronics parts in the car and the user of the car. It continually monitors the vehicle data and updates the server in real-time, which is then sent to the android application client. The main driver computer generates health reports based on the data acquired from the sensors. A constant equation usually derives this process. Figure 5.4 shows the connection between the electronics parts and the displaying unit in the car.

5.4   ON-BOARD DIAGNOSTICS

OBD systems play an essential role in the current generation of cars and play an increasingly important role in the next future. Parameters observed and diagnosed by the OBD system used by the ECU of the vehicle, which, in turn, controls the engine's main operation (like spark timing control, air-fuel mixture, fuel injectors spraying period, etc.) A combination of these two systems ensures safety and efficient vehicle operation. This unit is interfaced with vehicle sensors and reports the information and value obtained from each of the sensors. The ports and pin architecture for OBD-II is shown in Figure 5.5.

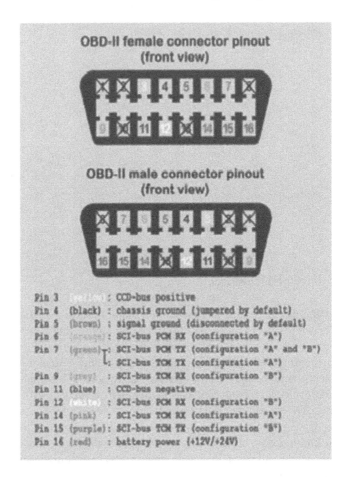

FIGURE 5.5   On-board diagnostic ports and pins.

5.4.1 Driver Assistance

The driver's main computer is a mini-computer as shown in Figure 5.6 that interfaced with the vehicle's ECU. The ODB interface performs communication with the vehicle dashboard and the engine ECU. Using the standardized scan codes, it collects real-time data from the vehicle dashboard and sends them to the driver's central computer. Sometimes this information can be helpful for the user or to the service personnel to help and diagnose the problem and to find out what causes the problem, enabling them to fix it easily and quickly.

5.4.2 Diagnostic Sensors

The driver's main computer and the OBD are interfaced with several sensors such as coolant temperature, pressure, air intake, fuel level, RPM, and ignition voltage sensors that are helping to identify the problem. Any abnormal value reflects a problem

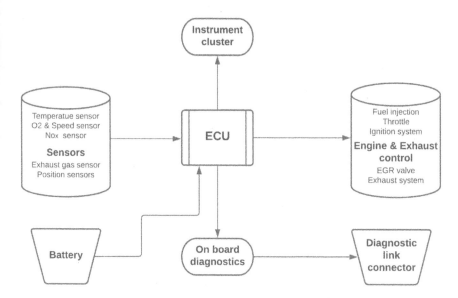

FIGURE 5.6   Interfacing of electronic parts.

that has already been occurred or yet to occur. Figure 5.7 shows interfacing different sensors with the micro-controller in our proposed system.

5.4.3   OBJECT DETECTION

Object detection can be done with the help of an open computer vision library by implementing it to the rear cameras in our vehicle. In our proposed work, by using the python programming language, this object detection mechanism is implemented more effectively. It alerts us whenever the obstacles are detected in our way. The frames from the rear camera are captured and sent to the micro-controller. Then micro-controller processes the frames and alerts the user if an object detected in the path. A display can also be implemented to help the user to view the objects in

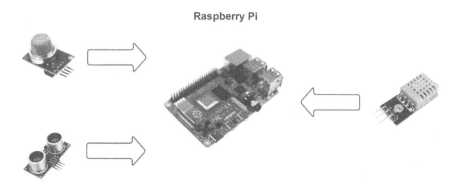

FIGURE 5.7   Interfacing sensors with micro-controller.

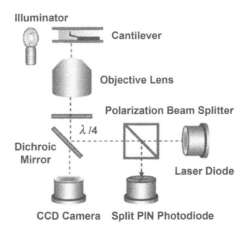

FIGURE 5.8 Camera position.

real-time, which is considered more convenient. The camera position in our car is illustrated in Figure 5.8.

5.4.4 SOFTWARE APPLICATION

The android application acted as a client-side unit and considered one of the significant modules in this system. It reports the sensor data which is monitored in real-time from the vehicle, as displayed in Figure 5.9. It is used for controlling the ignition

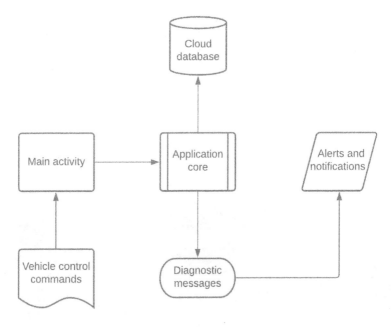

FIGURE 5.9 Android application architecture.

and air-conditioning system of the vehicle remotely. It usually obtains data from the server and also uses the server to send commands to the driver's central computer.

5.5 IMPLEMENTATION

OBD should be installed in the vehicle and interfaced with the data link connector. The Raspberry pi configured and installed as a computer inside the vehicle dashboard (Devi, R. S., Sivakumar, P., & Balaji, R. 2018). Users should download the smart vehicle application and register their vehicle in it for real-time monitoring and analysis. The vehicle is registered and identified using the vehicle identification number (VIN). The android application is used to report the data acquired from the vehicle, i.e., OBD connected to our mobile phones. The working principle of OBD is that the data sent to the server, and after loading, the data will be available back in our mobile and again sent to the server and forwarded to OBD. The hardware architecture is as shown in Figure 5.10.

Each DTC has a separate meaning, and each code defines a problem that has already occurred in the vehicle. Each code acts as a unique identification unit for a unique problem. The diagnostic code is created when a malfunction or misfire is detected. These codes help identify a problem quickly and effectively.

The first letter of the code indicates the family of DTC, i.e. Powertrain (gearbox, engine, transmission). The second bit indicates whether it is a manufactures or generic fault. The last three digits identify the type of problem as shown in Table 5.1.

5.5.1 INTERPRETATION OF RESULTS AND DISCUSSION

Figure 5.11 demonstrates the use of OBD after testing the car health with the help of our android application. This screen monitors the live sensor data, which is being

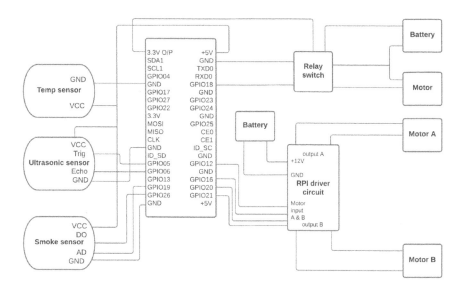

FIGURE 5.10 Hardware architecture.

TABLE 5.1
Trouble Codes

Codes	Description
P0300	Random/Multiple Cylinder Misfire Detected
P0654	Engine RPM Output Circuit Malfunction
P0700	Transmission Control System Malfunction
P0049	Turbo/Super Charger Turbine Over speed
P0093	Fuel System Leak Detected - Large Leak
P0350	Ignition Coil Primary/Secondary Circuit Malfunction
P0470	Exhaust Pressure Sensor Malfunction

reported by the vehicle in real-time. The data is provided by the main driver's computer, which was set up in the vehicle.

In the real-time simulation environment, the performance of identifying the exact trouble code occurs in the car is an important task. After identifying the trouble codes, it is an essential task for the user to select the best available channels to indicate the

FIGURE 5.11   Application results to the user.

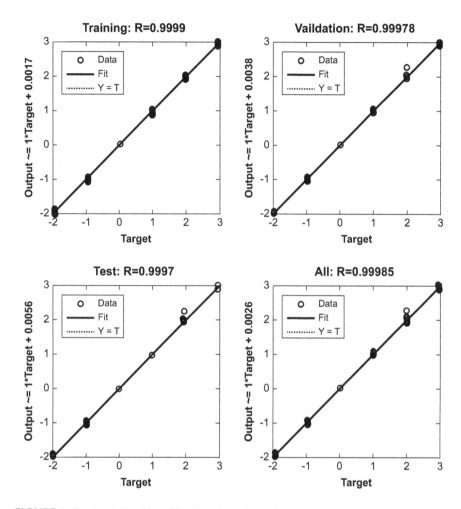

FIGURE 5.12   Analysis of identification of trouble code.

detected results. In this approach, input parameters such as speed, time, engine life, and accuracy have been considered, and the results are shown in Figure 5.12.

From Figure 5.13, it is found that the quality of identifying the trouble code has been improved. In training, it is clear that different combinations of input parameters have achieved 99% error-free results, during validation of input parameters, 99% of results have been completed. Then in testing the samples, it is found that 99% result is error-free. The overall effect of the Kriging artificial neuron network (ANN) model is 99%. Based on the simulation results, it is observed that the identification of exact trouble code probability is more accurate and maximum while using our simulation setup.

Figure 5.14 depicted that by using the artificial neural network, the error percentage was reduced, and the accuracy of identification of trouble code using the simulation setup is improved. By applying the different training and testing

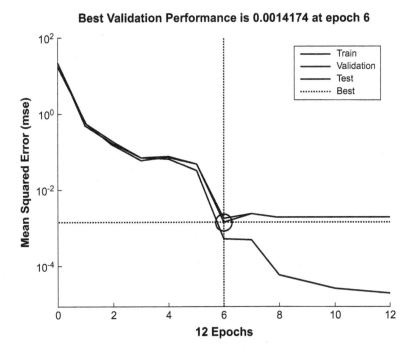

FIGURE 5.13   Validation of trouble code analysis.

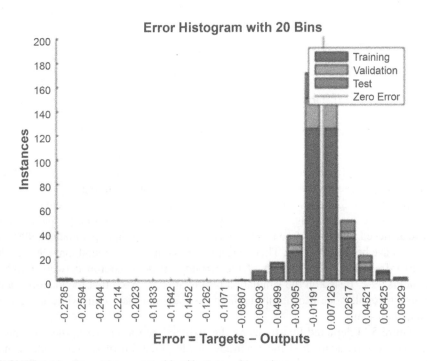

FIGURE 5.14   Error histogram to identify the trouble code.

cases, the efficiency of the trouble codes was more optimum by validating the concluded results.

The graphs show that the performance of OBD in a car has been obtained by keeping the input parameters and engine life as constant, whereas the parameters such as Time and Speed are varied. From that, it has been observed that the performance of OBD is high by keeping the time values more than the speed values. The graphs also show that the performance of OBD in a car is against the input parameters such as speed and time value is kept as constant, as shown in Figure 5.15.

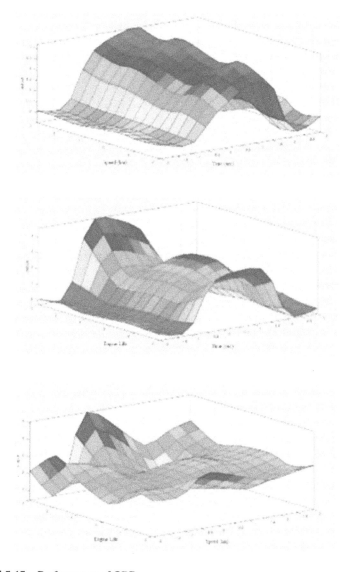

FIGURE 5.15   Performance of OBD.

5.6 CONCLUSION AND FUTURE WORK

The OBD system interfaces with the vehicle and the person who have been monitored and displayed in the mobile application. Problems are rectified and it will be displayed before arriving on the problem. The problems are identified through the sensor and the sensors send the information about the problems to the OBD system. The cloud server stores the data which is received by the OBDs system. The engine ignition and the air-conditioning are powered up with the relay circuits that are connected to the Raspberry pi. It can be started with the help of an android application when we are outside of our car. Future work is about to elaborate on the region radius. To make the region larger, the system should be upgradable to the wireless fidelity 6. To implement a dashboard camera that is capable of detecting lanes in the highway regions, it can be programmed with a more advanced open computer vision library. This might help in detecting driver sleep-over and may prevent accidents.

SUMMARY

It will be a complex task to produce vehicles with more assistance from the user's perspective. Since the vehicle does not have any portable device to communicate with the users and mechanism to clearly define the problem. We have proposed an OBD system that has software architecture to solve the issues and alert the users with the help of troubleshooting codes. The performance characteristics have been assessed through DTCs and the values were stored in the cloud through a mobile application. Based on the simulation results, it is observed that the identification of exact trouble code probability is more accurate and maximum.

REFERENCES

(Babu, M. S., Anirudh, T. A., & Jayashree, P. 2016) Babu, M. S., Anirudh, T. A., & Jayashree, P. 2016. Fuzzy system-based vehicle health monitoring and performance calibration. In *2016 International Conference on Electrical, Electronics, and Optimization Techniques (ICEEOT)* (pp. 2549–2554). IEEE.

(Devi, R. S., Sivakumar, P., & Balaji, R. 2018) Devi, R. S., Sivakumar, P., & Balaji, R. 2018. AUTOSAR architecture based kernel development for automotive application. In International Conference on Intelligent Data Communication Technologies and Internet of Things (pp. 911–919).

(Devi, R. S., Sivakumar, P., & Sukanya, M. 2018) Devi, R. S., Sivakumar, P., & Sukanya, M. 2018. Offline analysis of sensor can protocol logs without can/vector tool usage. *International Journal of Innovative Technology and Exploring Engineering*, 8(2S2) pp. 225–229.

(DiGiampaolo, E., & Martinelli, F. 2011) DiGiampaolo, E., & Martinelli, F. 2011. A passive UHF-RFID system for the localization of an indoor autonomous vehicle. *IEEE Transactions on Industrial Electronics*, 59(10), 3961–3970.

(Gallanti, M., Roncato, M., Stefanini, A., et al. 1989) Gallanti, M., Roncato, M., Stefanini, A., et al. 1989. A diagnostic algorithm based on models at different level of abstraction. *Variations*, 621(2), 4.

(Hu, J., Yan, F., Tian, J., et al. 2010) Hu, J., Yan, F., Tian, J., et al. 2010, March. Developing PC-based automobile diagnostic system based on OBD system. In *2010 Asia-Pacific Power and Energy Engineering Conference* (pp. 1–5). IEEE.

(Lu, W., Han, L. D., & Cherry, C. R. 2013) Lu, W., Han, L. D., & Cherry, C. R. 2013. Evaluation of vehicular communication networks in a car sharing system. *International Journal of Intelligent Transportation Systems Research, 11*(3), 113–119.

(Ruddle, A. R., Galarza, A., Sedano, B., et al. 2013) Ruddle, A. R., Galarza, A., Sedano, B., et al. 2013. Safety and failure analysis of electrical powertrain for fully electric vehicles and the development of a prognostic health monitoring system. In *IET Hybrid and Electric Vehicles Conference 2013 (HEVC 2013)* (pp. 1–6). IET.

(Sandhya, D. R., Sivakumar, P., & Balaji, R. 2019) Sandhya, D. R., Sivakumar, P., & Balaji, R. 2019. AUTOSAR architecture based kernel development for automotive application. In *International Conference on Intelligent Data Communication Technologies and Internet of Things (ICICI)*.

(Shiwu, L., Jingjing, T., Wencai, S., et al. 2011.) Shiwu, L., Jingjing, T., Wencai, S., et al. 2011. Research on method for real-time monitoring dynamic vehicle loading based on multi-sensors. In *2011 International Conference on Mechatronic Science, Electric Engineering and Computer (MEC)* (pp. 729–732). IEEE.

(Srihari, M. M., & Sivakumar, P. 2018.) Srihari, M. M., & Sivakumar, P. 2018. Implementation of Multi-Function Printer for Professional Institutions. In *2018 International Conference on Inventive Research in Computing Applications (ICIRCA)* (pp. 1–3). IEEE.

(Stefanowski, J. 2013) Stefanowski, J. 2013. Overlapping, rare examples and class decomposition in learning classifiers from imbalanced data. In *Emerging paradigms in machine learning* (pp. 277–306). Springer, Berlin, Germany.

(Sun, Q. 2002) Sun, Q. 2002. Sensor fusion for vehicle health monitoring and degradation detection. In Proceedings of the Fifth International Conference on Information Fusion. FUSION 2002 (Vol. 2, pp. 1422–1427). IEEE.

(Tunggal, T. P., Latif, A., & Iswanto. 2016) Tunggal, T. P., Latif, A., & Iswanto. 2016. Low-cost portable heart rate monitoring based on photoplethysmography and decision tree. In AIP Conference Proceedings (Vol. 1755, No. 1, p. 090004). AIP Publishing LLC.

(Zhang, Y., Salman, M., Subramania, H. S., et al. 2009) Zhang, Y., Salman, M., Subramania, H. S., et al. 2009. Remote vehicle state of health monitoring and its application to vehicle no-start prediction. In *2009 IEEE AUTOTESTCON* (pp. 88–93). IEEE.

6 Automotive Grade Linux
An Open-Source Architecture for Connected Cars

P. Sivakumar, A Neeraja Lakshmi, and
A. Angamuthu
Department of EEE, PSG College of Technology,
Coimbatore, India

R. S. Sandhya Devi
Department of EEE, Kumara Guru College of Technology,
Coimbatore, India

B. Vinoth Kumar
Department of IT, PSG College of Technology,
Coimbatore, India

S. Studener
Lilium GmbH, Weßling, Germany

CONTENTS

DOI: 10.1201/9781003269908-6

6.1 INTRODUCTION

The integrated work of a variety of technology results in a car. While each device is largely autonomous, the impact of other systems communicating with it has been affected by the recent developments in user behavior and consumer preferences (UX) that suggest a substantial need for modern graphical user interfaces (GUIs), which are often related and are more software-oriented, like an IVI framework. In order to survive in a dynamic market and develop the vehicle as a part of an automotive data value chain, original equipment manufacturers (OEMs) and Tier1 must live up to these demands (McKinsey & Company 2016).

A connected car is an automobile with internet connections and, in most cases, cellular connections. This provides cars access to information, data, applications and updates, connects with other IoT devices, and provides on-board Wi-Fi. By 2020, it is estimated that more than 381 million connected vehicles will be accessible on the road. This is a product of the car's durability. Like Tesla, General Motors (GM), Toyota, and BMW, the latter is now growing their related car portfolio.

At present, automobile manufacturers connect their cars in two ways: interconnected and tethered. Embedded vehicles have a built-in antenna and a chipset, while hardware is used to connect drivers' cars through their smartphones. Today, most of the car manufacturers use either Apple Carplay or Android Car to connect to the smart phones. The mobile phone attaches to the automobile, which can be controlled from an onscreen interface, from music calls to navigational directions on the phone.

The automated transfer of software, security updates, and online tracking, therefore, becomes important for OEMs, as consumer preferences are focused on the new smartphone characteristics. In reality, studies have shown that people today tend to buy a mobile rather than a far more expensive vehicle. In addition, as new connectivity services are required to deliver cost-efficient mobility, OEMs face a degrading market share in accordance with higher usage expectations. It is important for OEMs to standardize a generic software interface to handle vehicle costs during the design and development phases. The software of a vehicle provides the building blocks and can be changed to satisfy a range of consumer specifications to have more advantages. For example, several OEMs may standardize and execute the basic functions for the control of an electronic control unit (ECU). For car manufacturers, an appropriate operating system is expected to be an open-source method as well. The OEMs are, thus, helpful in the development and production process by providing a common software interface to handle vehicle costs.

The Linux Foundation released automatic grade Linux as an open-source project in 2012 to build and produce a universal Linux automotive software framework. As such, the project goes beyond IVI and seeks to involve telematics and instrument clusters over the long term. The concept is that this framework should serve as a reference framework for OEMs and manufacturers, by introducing new technology and designing a personalized brand user interfaces, to relate to and use their own company product (Sandhya Devi, R. S., Sivakumar, P., & Balaji, R. 2019). Since AGL began operations in the field of automobiles, tier 1 manufacturers, electronics and silicon vendors, AGL has obtained funding from several different businesses. Jaguar Land Rover, Nissan, and Toyota became the first car makers to become participants. Further participants include Renesas, Samsung, Fujitsu, Intel, and Texas Instruments. Today there are more than 50 members of the AGL workgroup.

6.2 BACKGROUND

Because of recent technological leaps forward, connected cars are becoming increasingly popular at progressively lower price points. Wireless networking of the fifth generation has opened new possibilities in terms of reliability and the requisite quality provision needed. Basically, the speed at which devices work has increased to the point that it is sufficiently safe to be used. As more companies move into the arena and competition increases, the sensors that are required for communication are becoming more widely manufactured and cheaper. Both Apple and Google are becoming involved; the former is creating ways to link their smartphone brands, the latter is committed to becoming part of an alliance of other organizations that want to make Android widely used in vehicles as well as working on a full operating system.

The realization of a connected car, thus, needs more digitization than a conventional vehicle and more device modules. Multiple device modules, including steering, vehicle networking, controls, video, IVI, drive logging, ADAS, and individual driving functions are included in the driving process. The extensive use of coding for a vehicle amounts to up to 100 million lines of codes (LoC) compared to with just six million lines of code in an aircraft. In addition, today's vehicles' theme-based costs are primarily dominated by software costs which already account for 40% (Charette 2009). Thus, a high level of software convergence is needed for active safety applications, including collision warning, car overhaul warning, and pre-crash sensing.

For the low-level layers in OSI model, standard communication systems including CANs, Flex Rays, and others can be classified as in-vehicle networks (Zeng, W., Khalid, M.A., & Chowdhury, S. 2016). Lately, efforts have been made to standardize vehicle networking's for higher performance and low-latency implementations, e.g., to support ADAS and other life-critical communication systems. The vehicle is connected to cloud providers through wireless providers including the LTE, 5G, IEEE 802.11 (WIFI), and IEEE 1609.x protocols. These new networks also include automotive Ethernet, AVB and time-sanitations networks (TSN). Considering the application layer, the adaptive AUTOSAR and AGL provide simple modules for establishing communication links with a connected car via RESTful API. Due to the huge installation of classical AUTOSAR parts, hybrid and adaptive AUTOSAR software modules can be mounted in the car (Devi, R. S., Sivakumar, P., & Balaji, R. 2018).

The advantages and weaknesses of different networks and describe all the strategies needed to improve quality of service (QoS). In addition, the development of a robust in-car connection scheme with fully separate networks and automotive gateways is protected by two classifications of automotive gateway (Zeng, W., Khalid, M.A., & Chowdhury, S. 2016). The proof-of -concept architecture retains Google Android automobile extensions that have options to merge extensibility and protection specifications, includes Google Android as an extension of the third-party framework. This paper results in unworthy programs being unable to reach capabilities of vehicles for the safe running of a vehicle (Macario, G., Torchiano, M., & Violante, M. 2009).

Several IVI application and open technologies, such as HTML5 and JavaScript, were addressed (Ostojic et al., 2016). It uses WebGL for progressive graphic effects. A Java-based architecture has been developed to enable the migration into the proposed application world of cluster and IVI applications. Integration of systems that utilizing various mobile electronic technologies in an automotive infotainment system. These apps are built into the user experience and the production cost is reduced. In Volkswagen and Audi vehicles, the control components refer to the control buttons in the lower part of the panel or to the control buttons on the edges, respectively (Huger, 2012).

The AUTOSAR consortium (Aust, 2018) supplies the vehicle ECUs standardized core functions (Devi, R. S., Sivakumar, P., & Balaji, R. 2018; Martínez-Fernández et al., 2015). Adaptive AUTOSAR is also mentioned and offers software components which will make it easier for connected vehicles to deploy a simplified platform. Ashwin et al. (2013) addresses the recognition of drivers using the fingerprint detection model. The primary aim is to enhance driver profiling and vehicle safety in connected vehicles. The model is based on the CNN and long short-term memory (RNN) RNN/LSTM which include data from smartphone sensors. The model is based on a variety of different types of networks.

6.3   MAIN FOCUS OF THE CHAPTER: CONNECTED CARS

A connected car is one that has its own internet connection as shown in Figure 6.1, usually via a wireless local area network (WLAN) that allows the car to share internet access and data with other devices inside and outside the vehicle. Internet service is typically connected to the network of the surrounding city. Many analysts say connected vehicles plays a major role in Smart city (Sivakumar, P., Nagaraju, R., Samanta, D., Sivaram, M., Hindia, M. N., & Amiri, I. S et al., 2020). Internet networks can have traffic warning networks, crashes, and other security warnings communicated between the vehicles (Noah et al., 2013). Connected cars provide a broader variety of communication possibilities than many other mobile vehicles. As well as allowing their users to obtain real-time access to all kinds of information, they can facilitate communication between the car and the dealership and warn the emergency services if you've been involved in an accident.

6.4   FUNCTIONAL FEATURES OF A CONNECTED CAR

Using a head panel, infotainment panel, dashboard device within a connected car, many features can be implemented.

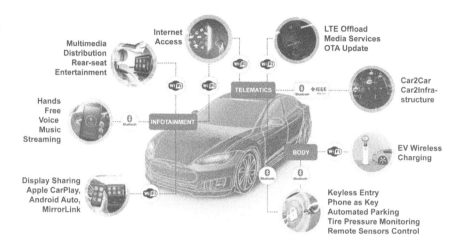

FIGURE 6.1   Connected cars structure.

6.4.1 Infotainment

1. Music/audio, podcasting, online radio, mobile, and online-enabled tablets for playing the music.
2. Smartphone/iOS/Android and proprietary games.
3. Bluetooth: This wireless link typically helps the driver to render and detach the phones from a physical device or tablet via Bluetooth. Microphones are also next to the driver so that the driver can communicate and the voice of the caller is played by the car speakers. Bluetooth can be used in streaming Bluetooth can be used in streaming Bluetooth wired music such as the iPhone, iPad, or 4G Wi-Fi hotspots.
4. Contextual assistance: The computer machine knows the needs of the driver in a new arena linked to the pipeline and delivers assistance, for example, when the gas tank is full, at gas stations.
   The voice commands are customized to the user as in the case, "I'm starving" is said by the driver, and nearby restaurants is seen.
5. Speech commands and hands-free controls: Speech commands like "Play my song" are frequently reactivated by the head device, and more advanced Siri alternatives such as "Sail to the nearest gas station" will become complex.

6.4.2 Navigation

1. Navigate either via a smartphone/iPhone app or via an optimized GPS navigation unit.
2. Leave reminders and SMS updates: The driver is able to get the right time to leave while linked to smartphones outside the car, while the car or smartphones advise friends or fellow drivers of the vehicle's arrival date.
3. Traffic on real-time: Advice on traffic and conditions in real time.
4. Dash payments: The opportunity to pay without leaving the car for products.

6.4.3 Safety Driving

1. Traffic, protection, crash warnings: The driver will be informed of car crashes, traffic delays, pot holes, or road debris when the car is linked to the GPS, city networks or group maps, such as Waze, HERE or vehicle-to-vehicle (V2V), vehicle-to-infrastructure (V2I), car-to-pedestrian/phone (V2P), or vehicle-to-everything (V2X) may be transmitted by the car in different manner to provide sophisticated details such as tapering or adjusting lights if no other vehicles are present at the same time (Ge, J. I., & Orosz, G. 2014; Subburaj et al., 2021).
2. Road side help: Certain services assist drivers by calling police or emergency personnel in case of collisions or other hazards. Studies also showed the prevention of accidents and the elimination of allegations by crash stopping devices.

6.4.4 Diagnostic Efficiency

1. Diagnostics for cars: Devices will alert the driver to the appropriate service and car issue.
2. The new function first from GM is predictive prognostics. Car owners can get warnings if the starter engine, fuel engine, or battery fails
3. Health records: Health reports are submitted to owners.
4. Apps for parking: Mobile parking software can find and pay for nearby car parking, such as parking systems, engine-style controls, and other ADAS functionalities.
5. Remote applications: Remote starting functions, remote door unlocks, remote A/C to heat or reset the vehicle, geo-fencing points, warnings, teenage and elderly surveillance and valet control. Remote applications (Guler, S. I., Menendez, M., & Meier, L. 2014).

6.5 CHALLENGES FACED BY AUTOMOTIVE INDUSTRY

Technologies like Smartphones and tablets have modified consumer perceptions. Their interfaces and the fast speed of innovation culminated in over 50% of customers requiring their cars to have similar features. However, automotive production varies considerably from smartphones and tablets with much longer production periods, tougher working environments, and more challenging systems with very little reuse of functionality between devices, much less between car manufacturers. Modern vehicles need more than 100 million lines of code between all their separate systems. Smartphone operating systems, such as Android, are managed on between 12 and 15 million lines of code, and a regular program ranges between 50,000 and 100,000 lines. The typical in-vehicle infotainment system needs between three and four years to develop, which usually takes three–four time cycles as compared to the development of smartphone.

Car manufactures intend to step up the production of in-vehicle operating systems to about the same pace as smartphones and tablets (6–12 months). The challenge of automakers and retailers as rapidly as mobile and tablet vendors in producing hardware and applications is fragmentation (Berret et al., 2017). The majority of mobile phones and tablets in the 2018 versions of the same operating system will run on Android, iOS, or KaiOS with a large backward compatibility. However, each automaker uses custom Linux, QNX, and Windows Automotive embedded models. Each car manufacturer not only uses customized variations of the operating system, but each car line in any car manufacturer uses various customized operating systems from different automobile vendors, even though they aim to standardize them where possible. The fast production cycles of smartphones and tablets make it extremely impossible for car manufactures to keep pace with the growth of car infotainment systems. The ideology and business principles are behind the technologies, along with the technology for designing vehicle infotainment system applications.

Car manufacturers choose to separate themselves from competitors, but don't enable more tech-centric firms who normally build operating systems, such as Apple and Google, to manage the user interface. While this is a legitimate problem, drivers hardly see several components of the car, as drivers hardly communicate with a vast variety of operating systems. The user interface is the primary competitive differentiator, thus, meeting the underlying functionalities in terms of speed, protection, and execution (Sivakumar et al., 2020). Any automotive producers want a proprietary operating system which will support the entire industry, while other car firms will want an open-source method. Business culture decides whether or not to create anything proprietary versus the use of open standards and business principles influences how much driver data are gathered, not to mention the processing and sharing of that knowledge with network partners.

6.6 SOLUTIONS AND RECOMMENDATIONS: AUTOMOTIVE GRADE LINUX (AGL)

AGL from the Linux Foundation hopes to be the mainstream in-vehicle information system in the de facto industry standard, including other technologies like advanced driver assistance systems, autonomous driving, the telematics market, and other in-vehicle displays. AGL plays an important role in addressing growth, fragmentation, and non-code reuse, among other factors in automotive industry. The Linux Foundation released Automatic Quality Linux as an open-source project in 2012 to build and produce a universal Linux automotive software framework. As such, the project goes beyond IVI and seeks to involve telematics and instrument clusters over the long term.

The aim is for OEMs and manufacturers to act as a reference forum to support and utilize their own consumer product by leveraging new software and designing a custom branded user interface. Since AGL began operations in the field of automobiles, tier 1 manufacturers, electronics, and silicon vendors, AGL has obtained funding from several different businesses. Jaguar Land Rover, Nissan, and Toyota became the first car makers to earn membership.

The goal of AGL is to promote the development and deployment of a truly open-source IVI vehicle computing framework (Marisetty S., Srivastava, D., & Hoffmann, J. A. 2010). In this context, AGL seeks to unite automakers and technology firms in the development and maintenance of a shared platform, at its heart, with Linux which enables OEMs to manage the user interface completely. This would provide a basis forum for OEMs, similar to GENIVI's strategy, which would concentrate their creative efforts instead of investing money in creating autonomous, individual solutions (Nathan, W., 2012; Swetha, S., & Sivakumar, P. 2021). Their aims, as themselves defined, are;

1. Provide an open-source Linux automotive platform that meets popular automotive needs.
2. Get an upstream distribution built for industrial goods to be modified and optimized.
3. Provide a guide delivery that illustrates the technology's advantages and strengths. Provide a software delivery to allow a quick start to engineering and easy prototyping experience.
4. Provide a large support group consisting of individual entrepreneurs, research institutions, and businesses.

The tasks are coordinated by the Linux Foundation, primarily by the Steering Committee, its coordinator and a series of expert groups. The above is in charge of designing work, while the steering committee sets the course of the project. The phase of implementation follows a "previous original" framework and all, individuals, businesses, and academic organizations, will contribute to the project.

6.7 STRUCTURE OF AGL

The structure of AGL is depicted in Figure 6.2. The development work of AGL is managed by expert groups who provide technical expertise and leadership by designing, implementing, testing, and documenting specific features of AGL. These are generally formed by the steering committee and are either responsible for a specific function (i.e., integration, quality assurance, management, etc.) or a specific part of the software (i.e., system maintainers for a particular code base). The expert groups are also responsible for collaborating with other open-source projects linked to AGL and to integrate appropriate solutions, as well as specification and evaluation of suitable hardware for the platform.

6.8 AGL TECHNOLOGY

The key role of the Linux Foundation in the project is to support AGL's vision by building an atmosphere of neutrality for cooperation and the development of working connections between OSS societies and the automotive sector. In addition, the Foundation offers guidance in the fields of academic, ethical, open source, and SPDX. The Steering Committee consists of a variety of organizations that strongly promotes the project's day-to-day operations. It encourages AGL's complete implementation and technological progress through support for protocols, standards, testing, and papers.

The basic building blocks of the AGL platform is quite similar to other embedded Linux platforms, as it is possible to use the same kernel and a lot of the same

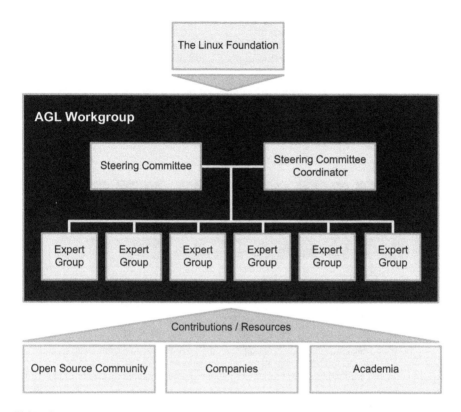

FIGURE 6.2   AGL structure.

middleware and open-source components. However, AGL has a higher interest in fast boot-time, security, and getting access to vehicle-specific buses such as the controller area network (CAN). The AGL platform supports a variety of different hardware architectures and is to some extent compliant with the GENIVI compliance specification (likely to be fully compliant in the near future).

Resource saving is very important in all creation and reuse of existing works in common. This refers to the development of open-source applications as well, which first offers a stable, better-tested framework, which AGL reacts to AGL. AGL can operate on a separate platform or start designing from scratch in other applications (e.g., telematics, instrument clusters, etc.).

Tizen is a Linux-based mobile operating system introduced by the Linux. The project was originally developed as an HTML5-based platform for mobile devices. Most of the software in Tizen is open source; it can be used as an operating system for IVI. Since Tizen IVI was originally developed for mobile and embedded devices as an IVI layer/profile for Tizen's operating system, the AGL IVI framework builds upon a commonly established and popular operating system platform with the addition of extra user interface and middleware components. Moreover, most AGL inventions are returned upstream to Tizen. Somewhere else in the middleware layer is the line between AGL and Tizen, as shown in Figure 6.3. The software architecture of

AGL can be represented with four layers beginning with the application on top and the HMI layer on top, followed by a system layer containing APIs for application creation and communication. The following is a layer of networks providing user-space facilities that are all program controlled. At the bottom, the kernel, various application drivers, and basic OS utilities are supported by a system layer.

The AGL framework's user interface and functionality, as shown in Figure 6.3, are made entirely of JavaScript and HTML5 and the framework can interact with

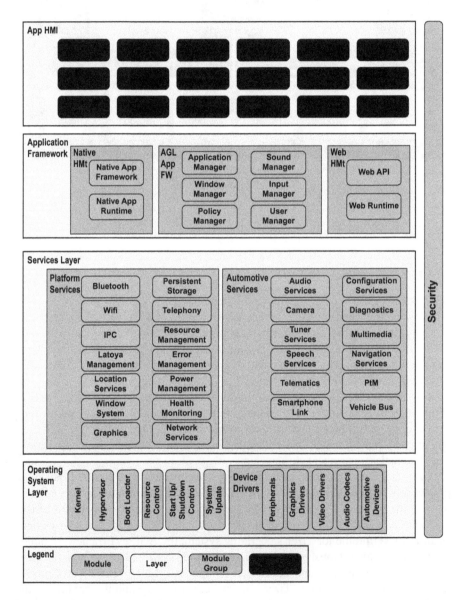

FIGURE 6.3   AGL framework.

the automobile using an automotive message broker (AMB) using the Tizen IVI web runtime in order to allow the application to pass data to and from the device. In turn, this is based on Crosswalk, an HTML5 framework runtime that expands the web platform to allow applications with specific runtimes to be deployed. The following list includes several features that are actually included in the AGL user experience.

1. Home screen.
2. Dashboard.
3. Air cooling and heating ventilation.
4. Google.
5. Media & news service.
6. Bluetooth services.
7. Audio services (MOST).
8. Connection from the near field (NFC).
9. Wireless internet connection.
10. Recognition of Fingerprint.
11. Recognition of voice.
12. Whether services.
13. Support for email.

AGL also includes other features like telematics, instrument cluster, ADAS system, autonomous driving, and so on, as shown in Figure 6.4 (Subburaj et al., 2021).

FIGURE 6.4   Supported features in AGL.

6.9   AGL'S UNIFIED CODE BASE

The AGL Unified Code Base (UCB) is a distribution focused on Linux that is set up by car manufacturers and vendors in collaborative attempts to provide users with the new infotainment for their vehicles and wired devices. The UCB infotainment framework is targeted at supplying 70%–80% of the development project's starting point. This encourages car producers and vendors to concentrate their efforts and tailor the other 20%–30% and their specific product requirements.

It includes main characteristics as shown in Figure 6.5 such as:

1. LGA structure for application.
2. ConnMan's multidevice network control.
3. Bus notifications for buses with optimized protection to avoid unintended intrusions.
4. Mixing and data routing.
5. Capability for different presentations (front and back seat).
6. Wi-Fi and LTE IP network admin.
7. Module for Linux security.
8. Yocto project Linux-based distribution.
9. Instrument cluster system profiles for telematics.
10. APIs for voice acceptance.

Many AGL participants have already begun the use of the UCB in their development plans. Subaru outback and Subaru Legacy 2020 uses open source AGL-UCB tech, Mercedes-Benz Vans uses AGL as the basis for a new online operating system for their commercial vehicles, and Toyota's infotainment system based on AGL is also used with Toyota and Lexus vehicles worldwide. The operating system, middleware,

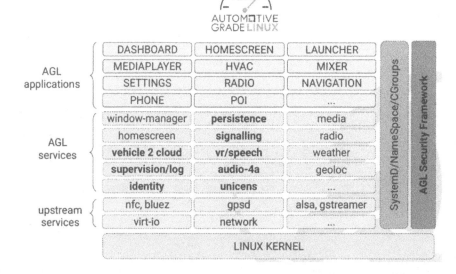

FIGURE 6.5   AGL unified code base structure.

and programmed stack are included in UCB 9.0/Itchy Icefish. Current AGL framework changes are:

1. Over-the-Air (OTA): Ostree update OTA.
2. Speech recognition: Alexa Auto SDK 2.0; Speech-API enhanced combining voice agents; latest open-source version of the voice recognition show card.
3. Application kernel: Enhancements including security implementation of Token Logic.
4. HTML5 software: Transformed to Token Logic protection, HTML5 picture accessible only with WAM and Chromium Web App Manager and HTML Sample Software for home-screen, launcher, dashboard, settings, media player, mixing device, HVAC and Chromium Browser applications.
5. Instruments cluster: QML comparison apps: LIN-to-IVI Functions, modified Instrument Cluster application with steering wheel/IVI Will messages.
6. Supporting board updates: Improved SanCloud BeagleBone Improved + Automotive Cape support Renesas RCar3 BSPs modified with v3.21 (M3/H3, e3, salvator), enhanced support for Raspberry Pi 4.

AGL-based infotainment framework is taking the AGL interface closer to the de facto mainstream in the industry. The market is just beginning to consider the value of open source and the possible effect of AGL on product growth. Toyota's presence is a big move forward; Toyota will potentially dramatically strengthen its market reputation and accomplishments along with other large car vendors, including Mazda, Suzuki, Honda, and Mercedes.

AGL does not provide commercial products in the open source. It's the foundation model instead. The starting point for a development project is between 70% and 80% AGL. Car producers add their own looks and sound for their own user interface, so it feels like their own company to install the applications they want.

6.10 INTEGRATION AMONG THE AUTOMOTIVE INDUSTRY

Car manufactures will personalize AGL to whatever they want. All AGL frameworks are the forum. It doesn't deal with the projection technology of smartphones like Android Car and Apple Car Play. You first need a robust framework for the screens to function in the vehicle. AGL will finally become the interface that manufacturers are most using, he stated. The mobile monitor doesn't replace device in the vehicle. However, it is still not available. The QNX owner is liable, but due to the popularity of Linux, they are fast losing market share. A major benefit for AGL is that a lot of those OEMs do not want a single-company system. You want your own life to be in charge. AGL helps you to tailor it to your own brand and do what you want.

6.11 DRIVING FACTORS

The purpose of the UCB infotainment framework is to include the most essential elements of the infotainment development framework. The sharing of a common software platform across the industry eliminates competition and accelerates time to the market by facilitating the development of a global community of developers that can construct a product once and work with several car manufacturers. The industry will start depending on the fact that AGL has a steady frequency of launches. Infotainment is ignored by AGL (AGL, "UCB"). The recently formed Virtualization Advisory Group is expected to take an active part in this transition as a virtualization framework and functionality can improve the security of the UCB and other functionality. It finds the UCB as a way to assist the UCB in other functionality including telematics, instruments, and heads-up displays.

6.12 VIRTUALIZATION IN AUTOMOTIVE

Today's cars hit the limits of their sophistication with hundreds of sensors, actuators, and ECUs that require communications of conjunction while guaranteeing highest efficiency, reliability, and protection. The automobile field will in future have an ever more complex driving and cutting-edge application for connected cars. In this setting, the number of applications rapidly grows in the hardware and components of the vehicle, resulting in an eruption of the architectural complexity of the vehicle (Aichouch, M., Prévotet, J.-C., & Nouvel, F. 2013).

In addition, increasing networking and applications contribute to the necessity of securing and defending a wider attack surface. The car business requires a hardware and software infrastructure that ensures isolation, streamlined systems management, high performance, transparency, interoperability, and stability to meet consumer requirements and include self-connected cars, always connected cars. An ECU equipped with the hardware components can run multiple execution environments with simultaneous, stable, and high-performance automotive functions. Examples of these attributes are:

1. Instrument cluster shows essential information (speed, signaling, etc.) and must be separated by reuse of the original software or regulations from the rest of the system.
2. IVI systems drive the multimedia/radio, ventilation and air conditioning (HVAC) central console, navigation, rear view camera, and many applications from third parties. Isolation is used as a protection precaution in the event that apps installed after the launch have potential vulnerabilities. In the instrument cluster, IVI systems will share display, video, and audio interfaces.
3. Telematics performs collection of telemetry data from vehicle, may also serve as OEM cloud connectivity gateway, and can support installation of edge services.
4. Safety critical functions include ADAS functions such as parking/lane assistant, autonomous driving, digital mirror, etc. These are typically implemented on top of AUTOSAR compliant operating systems (Sandhya Devi, R. S., Sivakumar, P., & Balaji, R. 2019).

6.13 AGL VIRTUALIZATION APPROACH

In order to be a de facto industry standard for automotive applications, AGL is creating a Linux-based, free computing platform. The AGL virtualization solution follows the same concept and is intended to provide a virtualization framework that can be used or replicated in a single hardware platform to combine multiple automotive functions. Virtualization is known as the next distinction driver in automotive applications. Virtualization is considered as a series of compatible systems that must be deployed, communicated, and interacted with.

Any car player can integrate these components to deliver personalized and exclusive solutions. AGL creates no new hypervisors, but uses latest open-source frameworks as its architecture modules. That is why the main pillars of the solution to AGL virtualization are:

1. Modularity: The AGL design is used as an integrated framework that can be linked throughout the compile period (and where necessary throughout run-time), hypervisors, virtual machines, AGL profiles, and automotive features. This differentiates through the mixture of components. Various modules can be instantiated. Modules may interact within themselves. Interoperability between open and proprietary modules is needed to achieve modularity.
2. Openness: There is no limitation on the use, implementation, and extension of the AGL virtualization platform. Variable hypervisors, CPU structures, device licenses that can be performed as a host or guest, are provided by the AGL virtual architecture.
3. Applications with various criticalities are aimed at co-existing and running digitally. The virtualization AGL solution, thus, seeks to combine software with various criteria for certification.

The virtualization solution as shown in Figure 6.6 is entirely consistent with existing AGL proposals and implementations, as is the AGL technology architecture. The AGL virtualization model introduced is also orthogonal. The AGL applications frame supports programmed separation based on namespaces, Cgroups, and

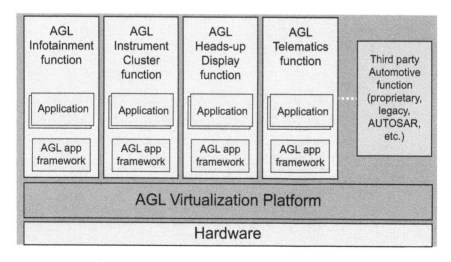

FIGURE 6.6   AGL virtualization hardware.

SMACK, which uses files/process protection attributes that any time an operation process is tested and which function well with protected booting techniques by the Linux kernel. However, where multiple systems are to be performed with various safety and security criteria (infotainment, instruments cluster, telematics, etc.), the control of these safety features becomes more complicated and an additional degree of separation is required to better separate these systems. This is the position of the AGL virtualization platform to improve device stability and separate numerous applications from AGL groups but even developers from third parties.

6.14 AGL SECURITY MECHANISMS

Security solutions included with AGL as seen in Figure 6.7, for example, Smack, Cynara, and AppFW may help to boost software application credibility. For global protection and separation of a complete device, access control [MAC/DAC] can be used.

1. Limiting users' permits: multiple users, such as the owner, mechanic, driver, and passengers, can be affiliated with a single vehicle. These users' freedoms are not equal and others may have more freedom. This is handled by account or profile permissions.
2. Restricting permits by request: An individual must ask for the requisite permits to work properly. The programmer will then access secured facilities or facilities if required approvals are issued.

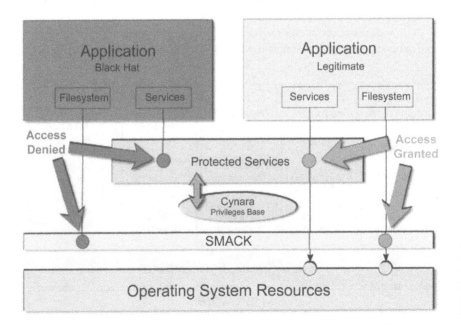

FIGURE 6.7 Security mechanisms in AGL using SMACK and Cynara.

6.15 ADVANTAGES OF AGL PLATFORM

6.15.1 COMMERCIALIZATION SUCCESS IN VARIOUS AUTOMOTIVE SECTORS

Toyota Camry's first AGL-based automotive infotainment interface launched in 2018 and rolled out to most North American Toyota and Lexus cars. Along with the first mass consumer deployment, AGL launched its UCB 5.0 in 2018 with a variety of hardware software applications for major automobile part manufacturers, including Intel, NXP, Qualcomm, Renesas, and Texas Instruments including reference apps for media, tuner, navigation, web server, Bluetooth, WIFI, HVAC control, audio mixers and vehicle control. While not participating in an AGL alliance, Tesla has developed a substantial portion of the open-source Model S and Model X apps and is already publishing the Linux source code for those projects in the middle of 2018. This is a testament to the open-source mechanism of adding technology back to the network of leading car producers, along with all the AGL participants operating as part of the Linux Foundation.

6.15.2 INNOVATIVE ENVIRONMENT

Nearly all competitive infotainment systems in the industry support automakers, and therefore, boost their market share through creativity and deliver attractive features at the best price. Infotainment systems on the market Innovation is at the forefront of AGL as part of a consortium. Automakers are able to reflect on what users actually want from their car infotainment experience rather than rebuilding underlying technology without driver experiences by working on noncompetitively segregated techniques. The key explanation for AGL is that this partnership is encouraged and creativity is supported where it really counts at the consumer touchpoint. Certain consortiums in the sector push diverse goals. The Free Automotive Partnership is committed to bringing Android into additional cars with a proprietary operating system and Google. The GENIVI Partnership has similar goals to AGL, but the operating system at higher level is agnostic to AGL, Android, and QNX by GENIVI.

6.15.3 DIVERSITY IN APPLICATION

AGL extends to telematics, instrument cluster, and heads up displays outside the in-vehicle infotainment market, with a potential roadmap for automated driver help, safety technologies, and ultimately stage 0–5 autonomous driving. Much just an operating system, AGL is indeed. It's a full computing stack including the Linux kernel, middleware, framework and APIs, SDKs, and test programs (Parekh, et al 2021). AGL provides a popular variety of service and interface in-vehicle infotainment products, offering standard front ends of another in-vehicle infotainments device, such as cameras and navigation systems, as a benefit of innovative firms GENIVI and the open automotive alliance.

6.15.4 CODE REUSABILITY

Technology from open source is between 60 and 80 percent new. These blocks are reusable components that allow developers to insert community-reliable and validated functions and features without the need to write any new code. They are elements which are reusable. AGL will then create a UCB that will be connected to other participating car manufacturers by developers while integrating its proprietary interface layers and data collection processes. This code sharing allows firms not to re-write underlying software to speed up their development process. Toyota states that the flexibility of AGL allows the in-vehicle infotainment system to roll out handheld and tablet usage systems across their cars more easily. Because of its open-source approach, AGL is focused on developing new features and marketing them more successfully, according to Toyota.

6.15.5 Price/Performance Value

Although there is no license expense for car producers and vendors for the underlying Linux technology, there is an on-going OPEX expense to support the program and contribute to the AGL project. The decision of the solution is of greater benefit to them, may be made by of automaker and infotainment supplier, but no upfront licensing provides a significant price/performance value advantage that must be taken into account in the infotainment system's total life cycle costs. The management of data management and consumer interaction over rival infotainment operating systems is a further factor for the price/performance benefit. AGL provides automakers with that option.

6.16 CONCLUSION

AGL is currently in its infancy, hitting its first commercial implementation in the 2018 Toyota Camry in a major OEM vehicle program. Toyota has revealed plans to roll out AGL over the coming years for the remainder of Toyota and Lexus cars. Such a high-profile implementation of mass-market cars will only support AGL as it becomes more popular, given that buyers respond well to the system. No other automobile operating systems have the supporting technology objectives of (1) acting as a de facto industry standard through the cooperation of producers, vendors, and providers of technology, and (2) encouraging creativity and enhancing time to market for in-vehicle infotainment by reducing proliferation of applications and reusing noncompetitive differentiated elements of the code base of the operating system. As manufacturers decide how best to integrate AGL into their infotainment systems and beyond, there has been continuous growth in the AGL frameworks.

REFERENCES

Aichouch, M., Prévotet, J.-C., & Nouvel, F. (2013). Evaluation of the overheads and latencies of a virtualized RTOS, In *8th IEEE International Symposium on Industrial Embedded Systems (SIES)*, pp. 81–84. doi: 10.1109/SIES.2013.6601475.

Ashwin, S., Loganathan, S., Kumar, S. S., & Sivakumar, P (2013). Prototype of a fingerprint based licensing system for driving, In *2013 International Conference on Information Communication and Embedded Systems*, pp. 974–987.

Aust, R. (2018). Paving the way for connected cars with adaptive AUTOSAR and AGL, In *2018 IEEE 43rd Conference on Local Computer Networks Workshops (LCN Workshops)*, pp. 53–58.

Berret, M., Mogge, F., Bodewig, M., Fellhauer, E., Sondermann, C., & Schmidt, M. (2017). Global Automotive Supplier Study 2018—Transformation in light of automotive disruption, Roland Berger and Lazard Automotive teams.

Charette, R. N. (2009, February). This Car Runs on Code. IEEE Spectrum. Retrieved from https://spectrum.ieee.org/transportation/systems/this-car-runs-on-code/.

Devi, R. S., Sivakumar, P., & Balaji, R. 2018. AUTOSAR architecture based kernel development for automotive application, In *International Conference on Intelligent Data Communication Technologies and Internet of Things*, pp. 911–919.

Ge, J. I., & Orosz, G. (2014). Dynamics of connected vehicle systems with delayed acceleration feedback. *Transportation Research Part C-emerging Technologies, 46*, 46–64.

Goodall, N. J., Smith, B. L., & Park, B. (2013). Traffic signal control with connected vehicles. *Transportation Research Record Journal of the Transportation Research Board, 2381*, 65–72. doi: 10.3141/2381-08.

Guler, S. I., Menendez, M., & Meier, L. (2014). Using connected vehicle technology to improve the efficiency of intersections. *Transportation Research Part C: Emerging Technologies, 46*, 121–131. doi: 10.1016/j.trc.2014.05.008.

Hüger, F. (2012). Platform independent applications for in-vehicle infotainment systems via integration of CE devices, In *2012 IEEE Second International Conference on Consumer Electronics - Berlin (ICCE-Berlin)*, pp. 221–222.

Macario, G., Torchiano, M., & Violante, M. (2009). An in-vehicle infotainment software architecture based on Google android, In *2009 IEEE International Symposium on Industrial Embedded Systems*, pp. 257–260.

Marisetty S., Srivastava, D., & Hoffmann, J. A. 2010. An architecture for in-vehicle infotainment systems. Dr. Dobb's. http://www.drdobbs.com/embedded-systems/anarchitecture-for-in-vehicle-infotainm/222600438 (Accessed by 12 June 2021)

Martínez-Fernández, S., Ayala, C., Franch, X., & Nakagawa, E. Y. (2015). A survey on the benefits and drawbacks of AUTOSAR, In *First International Workshop on Automotive Software Architecture (WASA)*, pp. 19–26.

McKinsey & Company. (2016). "Car data: paving the way to value-creating mobility—Perspectives on a new automotive business model," Advanced Industries.

Nathan, W. (2012). ALS: Automotive Grade Linux. LWN.net. Retrieved from http://lwn.net/Articles/517424/

Ostojic, R., Pesic, J., Bjelica, M., & Stupar, G. (2016). Java-based graphical user interface framework for In-Vehicle Infotainment units with WebGL support, In *IEEE 6th International Conference on Consumer Electronics - Berlin (ICCE-Berlin)*, pp. 184–186.

Parekh, T., Kumar, B. V., Maheswar, R., Sivakumar, P., Surendiran, B., & Aileni, R. M. (2021). Intelligent Transportation System in Smart City: A SWOT Analysis. In *Challenges and Solutions for Sustainable Smart City Development* (pp. 17–47). Springer, Cham

Sandhya Devi, R. S., Sivakumar, P., & Balaji, R. (2019). AUTOSAR architecture based kernel development for automotive application, In *International Conference on Intelligent Data Communication Technologies and Internet of Things (ICICI) 2018. ICICI 2018*. Lecture Notes on Data Engineering and Communications Technologies, vol 26. Springer.

Sivakumar, P., Devi, R. S., Lakshmi, A. N., Vinoth Kumar, B., & Vinod, B. (2020). Automotive grade Linux software architecture for automotive infotainment system, In *International Conference on Inventive Computation Technologies (ICICT)*, pp. 391–395.

Sivakumar, P., Nagaraju, R., Samanta, D., Sivaram, M., Hindia, M. N., & Amiri, I. S. 2020. A novel free space communication system using nonlinear InGaAsP microsystem resonators for enabling power-control toward smart cities. Wireless Networks, 26(4), 2317–2328.

Subburaj, S. D. R., Kumar, V. V., Sivakumar, P., Kumar, B. V., Srendiran, B., & Lakshmi, A. N. 2021. Fog and edge computing for automotive applications. In *Challenges and Solutions for Sustainable Smart City Development* (pp. 1–15). Springer, Cham.

Swetha, S., & Sivakumar, P. (2021). SSLA based traffic sign and lane detection for autonomous cars, In *2021 International Conference on Artificial Intelligence and Smart Systems (ICAIS)*, pp. 766–771.

Zeng, W., Khalid, M.A., & Chowdhury, S. (2016). In-vehicle networks outlook: Achievements and challenges. *IEEE Communications Surveys & Tutorials, 18*, 1552–1571.

7 Edge Node Creation Using Edge Computing Tools for Automotive Applications

P. Sivakumar, S. Bharanidharan, and A. Angamuthu
Department of EEE, PSG College of Technology, Coimbatore, India

R. S. Sandhya Devi
Department of EEE, Kumara Guru College of Technology, Coimbatore, India

B. Vinoth Kumar and S. K. Somasundaram
Department of IT, PSG College of Technology, Coimbatore, India

CONTENTS

DOI: 10.1201/9781003269908-7

SHORT SUMMARY

The Internet of Things (IoT) is the most recent, and it is expected to transport roughly 1.5 billion things globally—the IoTs might be a network of industrialized operating facilities. All of those devices produce an inexhaustible stream of data that must be saved and processed in real time for essential applications, a process that deployed cloud options will be unable to do. The most significant roadblocks to growth are slowing broadband rollout and data transmission delays among cloud servers and network clients. Edge computing solves these issues, indicating a paradigm shift in the cloud computing era. The majority of the data traffic produced by the Internet is currently carried by central data facilities. To provide an appropriate reaction time, data sources are now frequently transportable and located far from the main hub (delay).

7.1 INTRODUCTION

7.1.1 EDGE COMPUTING AND ITS USES

The Internet of Things (IoT) is the interconnection of hardware devices, automobiles, buildings, and other entities that are integrated with electronics, software, sensors, actuators, and networking links, allowing them to exchange data via the internet (Ai, Y., Peng, M., & Zhang, K. 2018).

With the rapid increase within the economy, the amount of transportation vehicles has grown extensively within the past few years. This results in a sequence of traffic-related issues, starting with the driver's and passenger's safety, hold up, and transportation delay. By intelligently connecting vehicles, the Internet of Vehicles (IoV) can give a better answer to the above issues (Harjula, E. et al 2019). As a result, each vehicle detects and transmits traffic-related data to other vehicles through the net. This helps in assisting both the drivers and passengers in providing autonomous driving, safe early learning, etc. The IoV provides clever and intuitive applications that use a lot of processing power and networking infrastructures.

Edge computing will play a significant role in analyzing and preserving information in geographically spread networks in this case (Ai, Y., Peng, M., & Zhang, K. 2018). The phrase "edge" refers to data that has been processed at the device's sting.

In the event of a processing resource in automobile applications, edge computing may be a better solution. Data collecting and assessment were delivered in close proximity to top computers, thanks to edge computing. Edge is a bridge between both the cloud and automobiles (Mahmud, R., Ramamohanarao, K., & Buyya, R. 2020) (El Zouka, H. A. 2016.).

7.2 NEED FOR EDGE COMPUTING

Currently, centralized data hubs form and process 91 percent of current data. Around 75% of most data will require investigation and intervention at the sting by 2022. Edge computing is more efficient for real-time analysis when compared to cloud and edge computing.

Edge computing provides higher quality of service (QoS) by distributing computing and warehousing resources around the client. Furthermore, a strong transmission and processing mechanism is necessary to support current applications in vehicular network (Hammoud, A. et al 2020).

The amount of data created on a continuous basis at the sting is expanding significantly larger than the ability of networks to analyze it. Endpoints should provide data to a grip computer, which processes or analyses it, rather than sending it to the cloud or a distant data hub to try to complete the work. The ultimate goal is to lower expenses and delays while maintaining network bandwidth control.

Edge computing also decreases the time it takes for data to be transmitted, processed, and then acted upon at the tip. Because plenty of data does not need to be transmitted up to the cloud to be processed and analyzed, assessment and incident processing may be done much faster and more effectively. Although cloud data hubs are several, if not hundreds, of miles away from interconnected items, round-trip latencies can range from tens to hundreds of milliseconds (Anawar, M. R. et al 2018).

The best way to minimize traffic jams and reduce traffic congestion is to use IoV-linked automobile solutions (Moregård, A. H., & Vandikas, K. 2015). The IoV is a set of automotive applications based on the IoTs that were created to meet the control goals of Intelligent Transportation Systems (ITS) (Asim, M. et al 2020) (Parekh, T. et al 2021). A wireless sensor network (WSN) is used in the IoV solution. Roadside units (RSUs) are fixed path points and service nodes that connect to network infrastructure via a single server-operated controller node. The WSN employs vehicle-to-vehicle (V2V) communication to transmit knowledge between multimodal perceptual interfaces and automobiles, and vehicle-to-infrastructure (V2I) transmission to send information between vehicle nodes and RSUs (Subburaj, S.D.R. et al 2021).

7.3 EDGE COMPUTING

7.3.1 GENERAL

The IoT is expected to carry roughly 50 million things worldwide by 2020 at the earliest—the IoT might be a network to major manufacturing facilities. All of those devices produce an inexhaustible stream of data that must be saved and processed in real time for essential applications, a process that established cloud solutions will be unable to do. The most significant roadblocks to growth are slowing broadband rollout and data transmission delays among cloud infrastructure and networked clients.

Public Cloud

Private Cloud /
Data-Center

Internet of Things

Edge-Gateway
Lokale Datenverarbeitung

FIGURE 7.1 An environment of edge computing.

Edge computing solves these issues, signaling a transition throughout the cloud computing paradigm. The schematic representation of an edge computing environment is shown in Figure 7.1.

7.3.2 Why Choose Edge Computing?

The majority of the data traffic delivered by the Internet is currently carried by centralized data hubs. To provide an appropriate reaction time (delay), datasets are now frequently mobile and located far from the main hub (Sandhya Devi R. S., Vijaykumar, V. R., & Sivakumar, P. 2021). Figure 7.2 depicts the architecture of more general edge computing (Raza, S. et al 2019).

7.4 EDGE IS A REPLACEMENT FOR THE CLOUD?

Edge computing is viewed as a supplement to cloud computing rather than a replacement. As a result, the edge serves a variety of following purposes:

Data gathering and collection: Edge calculation relies on data collecting near the source, while data sources in traditional cloud designs send all data to a critical cloud data center for central analysis. Microcontrollers are utilized directly on the device or on the gate—i.e., "intelligent routers"—for this purpose. These allow for the pre-processing and selection of data by integrating data from various devices. Only if the information cannot be reviewed locally, extensive analysis is necessary, or the information must be preserved and it is uploaded to the cloud.

Restricted data storage: Edge computing is very effective when local security of broadband transmission is required. Real-time data transmission from the primary data center to the cloud is usually unfeasible for big amounts of data. This issue can be avoided by keeping important data at the network's edge in a decentralized manner.

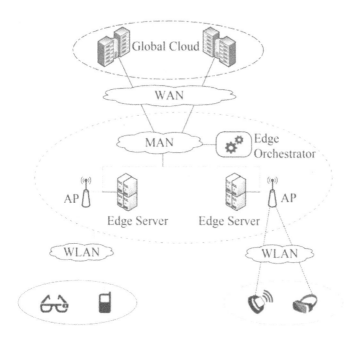

FIGURE 7.2   The architecture of general edge computing.

AI-monitored assistance: A peripheral computing environment's decentralized processing units collect data, evaluate it, and then monitor connected devices continuously. Real-time monitoring of status is achievable when machine learning methods are used.

7.5   SIGNIFICANCE OF EDGE COMPUTING

Edge computing is part of a larger ecosystem that could have unintended consequences. The following are some immediate significances:

- Improved speed and lower latency.
- Security.
- Cost savings.
- Remote reliability.
- Rapid scalability.

7.6   EDGE COMPUTING'S ROLE IN AUTOMOTIVE APPLICATIONS

Vehicle networks necessitate real-time storage and analysis of enormous amounts of data. Because of the high node latency and limited bandwidth in the communication channels, the information being shared is time sensitive (Sun, J. et al 2019). The network must also be aware of the location of the context. Up to 960 million connected vehicles are expected to be in operation by 2030, processing massive amounts of data on a daily basis. For a technically possible solution that meets the QoS and quality of performance (QoP) standards, the communication mechanism must be resilient.

Fog and edge computing (FEC) frameworks provide IoT systems with additional encryption in order to ensure transaction confidentiality and trust. For example, to address a security issue, today's wireless sensors put in outdoor situations usually require remote upgrading of wireless ASCII text files. However, due to a variety of environmental circumstances such as fluctuating signal strength, interruptions, and bandwidth constraints, the remote-centralized backend server may struggle to complete the update in a timely manner, raising the risk of a vulnerability attack.

If the FEC framework is used to merely update the devices security for the wireless sensors, the backend would establish the quickest way through the network across multiple nodes. The FEC assists them in considering the goals of their customers by encouraging autonomous decision-making in the areas of compute, storage, and control. As a result of FEC's understanding of numerous self-adjustments, self-organization, subconscious, actualization, and other processes, IoT devices are transformed from passive to smart active devices that can continuously run and respond to consumer needs without relying on cloud-based decision making.

7.7   EDGE COMPUTING IN THE VEHICLE: SEVEN USE CASES

Edge computing offers a variety of techniques for creative applications and enhancements to the user experience within the car, and its capabilities go far beyond this one-use scenario. Edge computing is concerned with data generated close to its source, such as by a smart watch, an industrial robot, or a vehicle. As a result, the word complements cloud computing, which refers to the processing capacity available in data hubs. However, in terms of processing power and memory, the above-mentioned intelligent gadgets do not appear to be as capable as computers in data hubs. PORSCHE Newsroom presents seven scenarios in which edge computing can be a big help in the car.

Sensor fusion and value aggregation: Many sensors are found in commonplace electronics. There are many different sensor kinds, form-factors, and models available, ranging from gyro sensors in mobile phones to humidity sensors in a smart home. They all have one thing in common: they can produce a lot of data. This also holds true for cars with a large number of sensors. Although the majority of the data will be processed on-board, some applications, such as warning in the instance of aberrations from the norm, will necessitate data transfer to the cloud. Edge computing helps to limit the amount of data that is sent in a smart manner, lowering data transmission costs while also lowering the amount of sensitive data that leaves the vehicle.

Independent driving and intelligent communications: Autonomous driving is a unique case of edge computing because the driving algorithms must be conducted in real time within the vehicle's control unit. Another application of the phrase "fat edge" is in smart infrastructure, such as 5G base stations, intermediary smart routers, or traffic lights. This allows for substantially improved efficiency and throughput at junctions, for example. As an example, consider a fancy and heavily used crossroads with five lanes and long traffic light wait times for every vehicle. Because it follows the timing of traffic signals, autonomous driving alone would not eliminate waiting times. However, if a footing node is built at the intersection, the vehicles' trajectories will be received from the sting node as they approach the intersection. Rather of computing each car separately, this edge node may orchestrate all surrounding vehicles.

Intuitive promotion with machine learning: Aside from the driving controls, an extremely vehicle's entertainment system could be a very prominent computer application. Machine learning algorithms are a critical tool for finding useful insights into enormous amounts of data to determine what functionalities and applications people are actually utilizing and where interaction design could be maximized—whether it's the touch or voice interface. Edge computing facilitates the deployment of cloud-trained machine learning models on mobile devices. In order to improve user interaction, local behavioral and sensor data are used to make predictions. Because the system learns the behavior or some environmental limitations, it is envisaged that interaction in the post-personal assistant period will become simpler over time.

Flexible projecting protection on electric vehicle: Because Porsche is known for its high levels of driver involvement and performance, the batteries in our electric vehicles must match this. Battery monitoring and predictive maintenance are important components in achieving this goal. Tire pressure, acceleration, traffic, charge cycles, driver behaviors, and other dynamic events affect battery maintenance and charging after the car has been shipped out. A viable solution would also take into account user data, which isn't available at the time of manufacture, and could then, for example, modify the suspension to the drivers' personality. With the ability to aggregate data and near real-time evaluation of relevant battery metrics and sensor outputs, edge computing can help.

Multifactor authentication for an easy: This case is about delivering a frictionless access to the car that is supported by numerous security factors, allowing consumers to enter their car's drivers' door without any friction. This might be accomplished by employing (1) a camera for face identification, (2) an infrared camera for impersonating recognition, and (3) a Bluetooth sensor to determine the drivers' transportable vicinity. These three factors are examples of multifactor driver authentication.

Integration of smart houses and vehicles: There are few circumstances where computational power is required at the point of advancing smart homes and Porsche as a luxury brand. For example, a valet to house service may be available; allowing the customer to leave their automobile parked in their driveway and just goes. Within the garage, the automobile will park itself. The automobile can also drive itself out of the garage and prepare for a replacement start, according on the owner's personal calendar schedule. However, autonomous driving skills are required to make this use possible.

The Car's screening and notification: Car network operators will have to administer infrastructure in connection to their vehicle, which will necessitate a lot of extra monitoring and restrictions. As an example of a rule like this: "Don't dispatch this car because it's up for service within the next 50 miles if a customer is booking it for a four days trip—meaning that it's very likely to travel over the 50 miles left." Without using the cloud in any way, edge computing can analyze accessible sensor data and assess principles and machine learning techniques immediately within the car. Cloud computing is frequently used to train and develop machine learning algorithms that are required to recognize specific scenarios. In addition, anytime an update is produced, the foundations definition will be drained from the cloud and pushed all the way down to the sting.

7.8   APPROACH ON EDGE COMPUTING

The edge server environment is constructed using node.js and Angular 7. Node.js is a fully accessible server environment capable of creating, opening, reading, writing, closing, and deleting files on the server. When a node is in this environment and is within range of the server, the node is automatically connected to the server. Wi-Fi is used to connect the nodes and the server. From then, the server and nodes are connected with one another by sharing and processing data.

7.8.1   ANGULAR 7

Angular 7 is a fully accessible JavaScript paradigm for creating web apps and apps in HTML, JavaScript, and Typescript, which is a basic structure of JavaScript. Nowadays, Angular is the most reliable and popular JavaScript supported solution. Environment setup required for Angular 7. To install Angular 7, we require the following:

- Nodejs.
- Npm.
- Angular CLI.
- IDE for writing code.

7.8.2   NODE.JS

- Node.js is a fully accessible server environment.
- Node.js is a free programming language that may be used on a variety of platforms (Windows, Linux, Unix, Mac OS X, etc.).
- On the server, Node.js employs JavaScript.

Opening a file on the server and returning the content to the client is a common activity for an internet server.

This is how Node.js deals with a file request:

- Able to handle the following request by sending the work to the device's classifying process.
- The server responds the information to the client after the classification system has opened and skimmed the file.

Node.js minimizes the wait time and allows you to quickly go on to the next request. Node.js allows for single-threaded, nonblocking, asynchronous programming, which saves a lot of memory.

What does Node.js have to offer?

- Node.js has the capability of generating the page content more dynamically.
- On the server, Node.js can create, load, access, edit, remove, and close files.
- Node.js is capable of gathering data from forms.
- Node.js has the ability to create, remove, and alter data in your database.

7.8.3 VIRTUAL STUDIO CODE

Visual studio code is a small but capable ASCII text file editor that runs on your computer and is available for Windows and other operating systems as well. It includes built-in support for JavaScript and Node.js, as well as a vast environment of extensions for many other programming languages (including C++, Java, C#, Python, and PHP) and runtimes (such as .NET and Unity tools).

7.9 WORKING ENVIRONMENT FOR EDGE COMPUTING

The edge server environment is constructed using node.js and Angular 7. Node.js is a fully accessible server environment capable of creating, opening, reading, writing, closing, and deleting files on the server and version of the installation packages as shown in Figure 7.3. By using this environment, a node in proximity to the server range, the node gets automatically connected to the server. The nodes and server are connected through Wi-Fi. From then, the server and the nodes are connected with one another by sharing and processing data. The data (Speed) has been sent to the server manually, it processes the data in a server environment. Based on the speed read in the server, it creates an impact on the node whether it is in threshold speed.

FIGURE 7.3 A version of the installation packages.

7.9.1 Procedure to Approach

- Check the installation packages for the edge environment. From the command prompt create a workspace for the project.
- After creating the environment open virtual studio IDE. The default packages of the modules are already created in the file.
- In that package, open the .html file and write the code for front end platform.
- In app.component.ts, code for processing data should be written.
- In command, prompt compile the file by using the command ≫ng serve.
- After compilation, it creates the local host Id. 192.168.225.117:4200.
- When a node is within the range of server, it get automatically connected to the server.

7.9.2 OUTPUT

As seen in Figure 7.4, Data from IoT devices can be examined at the network's edge before being delivered to a data center or cloud, thanks to edge computing. The speed is monitored in the car, and the data is transferred to a nearby data center. In the particular local data center, the average speed of the car in the specific place has been processed. As a result, all of the automobiles are linked to the internet. Data can be sent to vehicles from a data center while attached to the internet. The vehicle's average speed in a specific location is relayed to the automobile to warn of rash driving and avert an accident.

The nodes is connected to the data center, the speed has been sent to the server manually. In a server, the data has been processed and revert to the node. Here server is the laptop and the nodes are mobiles phones (Mao, Y. et al 2017). The mobile phone having the toolbox to change the speed is shown in Figure 7.5

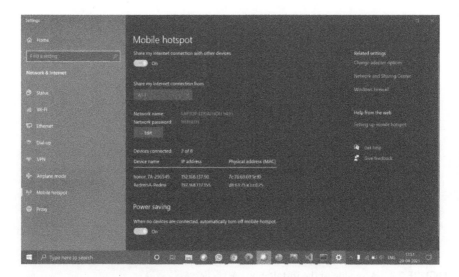

FIGURE 7.4   The node is connected to server.

FIGURE 7.5 Speed changes in the nodes 1 & 2.

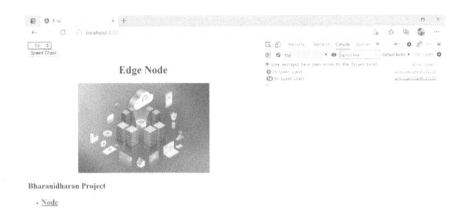

FIGURE 7.6   Console process of output.

Let see how the speed process is done in the console in Figure 7.6. If the speed is within the speed limit, the console shows it is "in speed limit," and if exceeds the average speed, it will show "Maintain within Speed Limit."

7.10   REVIEW OF THE WORK DONE

The impact of Edge computing in automobile's contexts is deliberated in this chapter. Within each of these elements, it provides a full overview of the theory, organization, techniques, and uses. Cloud computing aids the IoT infrastructure for intelligent vehicles in a variety of ways, but fog and edge solutions are unavoidable solutions to several of the IoT concerns that arise in the vehicular settings. The data (Speed) has been sent to the server manually, it processes the data in a server environment. Based on the speed read in the server, it creates an impact on the node whether it is in threshold speed. The output is simulated through Angular 7, Node.js.

7.11   CONCLUSION

The data that has been sent to the server is done manually, but for future use, by using any sensor the data can be sent automatically. To ensure precision, contact-based systems that leverage on perceptual network administration and nonmethods are used to identify vehicle congestion and improve road safety. IoV-based solutions are well-suited to applications that require high visual data fidelity and are time-sensitive. The accuracy rates of V2V and V2I traffic concentration estimations are within tolerable bounds. Low turnaround time, low connection latency, portability support and position awareness are all requirements for IoV systems that collect significant volumes of data for interpretation. Reliant solutions for VEC and Vehicular Fog Computing (VFC) meet the preceding objectives. However, the VFC-reliant approach makes use of minimal algorithms and simplified application interfaces to improve network turnaround time and handle a wider range of apps. For applications requiring high sensitivity vehicle tracking and capacity, the fog computing centric approach with portability and integrated architecture can be beneficial.

REFERENCES

Ai, Y., Peng, M., & Zhang, K. 2018. Edge cloud computing technologies for internet of things: A primer. Digital Communications and Networks. 4(2), 77–86. https://doi.org/10.1016/j.dcan.2017.07.001.

Anawar, M. R., Wang, S., Azam Zia, M., Jadoon, A. K., Akram, U., & Raza, S. 2018. Fog computing: An overview of big IoT data analytics, Wireless Communications and Mobile Computing, 2018, 7157192.

Asim, M., Wang, Y., Wang, K., & Huang, P. Q. 2020. A review on computational intelligence techniques in cloud and edge computing. IEEE Transactions on Emerging Topics in Computational Intelligence, 4(6), 742–763.

El Zouka, H. A. 2016. A secure interactive architecture for vehicular cloud environment. In 2016 IEEE International Conference on Smart Cloud (SmartCloud) (pp. 254–261). IEEE.

Harjula, E., Karhula, P., Islam, J., Leppänen, T., Manzoor, A., Liyanage, M., … Ylianttila, M. 2019. Decentralized IOT edge nanoservice architecture for future gadget-free computing. IEEE Access, 7, 119856–119872.

Hammoud, A., Sami, H., Mourad, A., Otrok, H., Mizouni, R., & Bentahar, J. 2020. AI, blockchain, and vehicular edge computing for smart and secure IoV: Challenges and directions. IEEE Internet of Things Magazine, 3(2), 68–73.

Mahmud, R., Ramamohanarao, K., & Buyya, R. 2020. Application management in fog computing environments: A taxonomy, review and future directions. ACM Computing Surveys (CSUR), 53(4), 1–43.

Mao, Y., You, C., Zhang, J., Huang, K., & Letaief, K. B. 2017. A survey on mobile edge computing: The communication perspective. IEEE Communications Surveys & Tutorials, 19(4), 2322–2358.

Moregård, A. H., & Vandikas, K. 2015. Computations on the edge in the internet of things. In 6th International Conference on Ambient Systems, Networks and Technologies (ANT)/5th International Conference on Sustainable Energy Information Technology (SEIT) (pp. 21–26).

Parekh, T., Kumar, B. V., Maheswar, R., Sivakumar, P., Surendiran, B., & Aileni, R. M. (2021). Intelligent Transportation System in Smart City: A SWOT Analysis. In Challenges and Solutions for Sustainable Smart City Development (pp. 17–47). Springer, Cham.

Raza, S., Wang, S., Ahmed, M., & Anwar, M. R. 2019. A survey on vehicular edge computing: architecture, applications, technical issues, and future directions. Wireless Communications and Mobile Computing, 2019, 3159762. https://doi.org/10.1155/2019/3159762

Sandhya Devi R. S., Vijaykumar, V. R., & Sivakumar, P. 2021. Edge Architecture Integration of Technologies. In Cases on Edge Computing and Analytics (pp. 1–30). IGI Global.

Subburaj, S. D. R., Kumar, V. V., Sivakumar, P., Kumar, B. V., Surendiran, B., & Lakshmi, A. N. 2021. Fog and Edge Computing for Automotive Applications. In Challenges and Solutions for Sustainable Smart City Development (pp. 1–15). Springer, Cham.

Sun, J., Gu, Q., Zheng, T., Dong, P., & Qin, Y. 2019. Joint communication and computing resource allocation in vehicular edge computing, International Journal of Distributed Sensor Networks, 15(3), 1550147719837859.

8 Nanosensors for Automotive Applications

M. Saravanan
Department of ECE, Sri Eshwar College of Engineering,
Coimbatore, India

E. Parthasarathy
Department of ECE, SRM Institute of Science and Technology,
Chennai, India

J. Ajayan
Department of ECE, SR University, Warangal, India

P. Mohankumar
Department of Mechatronics, Sona College of Technology,
Salem, India

CONTENTS

DOI: 10.1201/9781003269908-8

8.1 INTRODUCTION

The use of onboard sensors combined with internet connectivity offers a very good driving experience, which propels the rapid growth of automotive market. The ever-growing sales of automotive across the world and emergence of new automobile technologies such as autonomous vehicles are the key factors driving the growth of automotive sensor market. It is expected that nanotechnology is going to play a very significant role in the automobile industry. Nanosensors can be used to sense emission gases like hydrogen, oxygen, carbon monoxide, carbon dioxide, and nitrogen oxide in vehicles. Nanosensors can also be used for monitoring the temperature and pressure in automobiles. This chapter deals with nanomaterials and nanostructures used for the development of nanosensors for automotive applications. This chapter also throws lights on the use of graphene and carbon nanotubes (CNTs) for gas sensing, nanoparticle-based pressure sensors and temperature sensors, silicon and other semiconductor-based pressure and temperature sensors, metal oxide gas sensors and application of nanotechnology and nanostructured materials in automotive applications. The demand for nanosensors have been increasing in the automobile industry because of the environmental requirements, features that provide more comfortness, electronic stability programs (ESP) and their requirements in safety features like airbags. The future for automotive sensor business seems to be bright due to the increasing safety and comfortness based demands of customers.

8.2 APPLICATION OF NANOTECHNOLOGY IN AUTOMOTIVE INDUSTRY

Nanotechnology is expected to play a crucial role in the automobile industry in the future because nanomaterials and nanostructures can provide better performance compared with existing materials and technologies. Nanotechnology can be applied to tires and chassis, emissions, body parts, drive trains and engines, sensors, and other electronics and interior of the automotive. The potential applications of nanotechnology in automobile sector are

- Nanomaterials in battery/fuel cells.
- Nanosensors.
- Nanoparticles in tires.
- Catalyst nanomaterials.
- Nanograined body/engine.
- Polymer nanocomposites in body parts.
- Nanocoating in vehicles.
- Nanomaterials in lubricants.
- Nano additives in fuel.
- Nano polymer materials for vehicle body glancing.
- Nanomaterials and structures for heat shield.
- Nanofilms for self-cleaning.
- LED displays.
- Antifogging and antireflective coatings.
- Solar cells in body parts.

8.3 NANOPARTICLES AND NANOSTRUCTURES FOR AUTOMOTIVE APPLICATIONS

Nanotechnology is considered as one of the major technologies for the next-generation automotive industry to maintain competitiveness. Nanomaterials and structures in automotive industry provide benefits like corrosion resistance and reduced wear and tear, resistance against UV rays, development of nanosensors and other electronics, less weight, and minimization of emissions. It is expected that, in 2030, the number of vehicles in the world will reach 1.5 billion. Nanostructured materials such as CNTs, graphene, nanotubes, nanofibers, nanorods, nanowires (NW), nanoparticles, and nanosheets can be used for the development of area, cost, and energy efficient nanosensors for the future automotive industry. Apart from these materials, solid-state semiconductors such as silicon (Si), germanium (Ge), GaN, GaAs, InGaAs, InAs, InAlAs, InGaN, etc., can also be used for developing nanosensors for automotive applications. The automotive sensor business seems to have an extremely good future due to the growing demand for comfortness and safety. Nanosensors in automobiles also ensure the emission friendliness, fuel efficiency, and safety of vehicles.

8.4 NANOSENSORS FOR GAS SENSING

Nano gas sensors are expected to play a very big role in next-generation combustion-based vehicles to monitor exhaust gases. High dynamic flow rate and high gas temperature are the key challenges in the design of nanosensors for exhaust gas detection especially in diesel engines (H. Zhang, J. Wang, Y.-Y. Wang, 2015). Automotive exhausts consist of highly poisonous gases like NO_x, CO, CO_2, SO_2, and H_2. Zirconia-based gas sensors are considered as the most suitable gas sensors due to their unique characteristics such as mechanical and thermal robustness (T. Ritter et al, 2018). Thermoelectric hydrocarbon gas sensors are found to be highly effective in harsh conditions (G. Hagen, A. Harsch, R. Moos, 2018). Thermoelectric hydrocarbon sensor consists of an inert area and a thick porous platinum (Pt) filled alumina film-based catalytically activated area. Let ΔT be the temperature difference between these two areas. A thermopile fabricated using thick film technology can be used for measuring ΔT. The use of low thermal conductivity substrate materials and large number of thermopile structures are the two most effective ways to improve the gas sensing ability of thermoelectric hydrocarbon gas sensor. Pt and Au are widely used as thermopile materials. The sensor signal depends only on the exothermic heat resulted by the oxidation of gas at the activated region. CP, MOS, and MOSFET sensors are the first-generation gas sensors and MS-chemo sensors and quartz crystal microbalance are the second-generation gas sensors used in the automobiles (S. Garrigues, T. Talou, D. Nesa, 2004). The gas sensor directly converts the concentration of gas into an output electrical signal.

8.4.1 NO_x NANO GAS SENSORS

NO_x is a common and unavoidable emission in petrol, diesel, or gas engine-based vehicles. O_2 concentration, temperature, and reaction duration are the key parameters that determines the emission rate of NO_x (S. Saponara et al, 2011). The thick film technologies like screen printing can be used for manufacturing β-alumina ($2Na_2O$-$11AL_2O_3$) protected gas sensors that can be used for sensing NO_2, CO, and hydrocarbon (E. Billi et al, 2002). The detection range of β-alumina sensor is 10–1000 ppm. This sensor works based on the mechanism of chemisorption of O_2 that creates a capacitance effect at the electrolyte-metal interface which produces an output voltage signal proportional to temperature, type, and concentration of gases. Na^+ ionic conductor and β-alumina forms the solid-state electrolyte of the sensor and Pt and Au forms the metal electrodes for the device. The metal electrodes are in contact with the gas. Sol-gel process can be used for preparing β-alumina powder. Sputtering process can be used for depositing Pt and Au metal contacts on pellet surface (N. Guillet et al, 2002).

Yttria-Stabilized Zirconia (YSZ) and Scandia Stabilized Zirconia (SSZ) materials can also be used as a sensing element for NO_2 detection sensors and asymmetric, semicircular and inter digitated geometries can be used for designing sensing elements (D.L. West, F.C. Montgomery, T.R. Armstrong, 2005). The output voltage between the electrodes of a zirconia-based gas sensor can be computed using Nernst's Law (S. Zhuiykov, N. Miura, 2007).

$$V = \overline{t_i} \frac{RT}{2nF} \ln \frac{P_{O_2}(\text{gas})}{P_{O_2}(\text{reference})} \tag{8.1}$$

where
  V = Output voltage
  $\overline{t_i}$ = average ionic transference number
  R = Universal gas constant
  F = Faraday's constant
  T = Absolute temperature in K
  P_{O_2} (gas) = O_2 partial pressure at sensing element
  P_{O_2} (reference) = O_2 partial pressure at reference element.

WO_3, $NiCr_2O_4$, $ZnCr_2O_4$, $ZnFe_2O_4$, ZnO, Cr_2O_3, NiO, $LaFeO_3$, $La_{0.8}Sr_{0.2}FeO_3$, $La_{0.85}Sr_{0.15}CrO_3$/Pt, ITO (tin doped indium), $La_{0.6}Sr_{0.4}Fe_{0.8}Co_{0.2}O_3$ and $CuO+CuCr_2O_4$ can be used as sensing element materials. The zirconia-based sensor output as a function of NO concentration is depicted in Figure 8.1. The sensor output is directly proportional to NO concentration. J.C. Yang and P.K. Dutta (2007) demonstrated a high-temperature electrochemical YSZ NO_x sensor that featured a Pt inserted zeolite Y-filter. The output signal is found to be decreasing with increase in Pt inserted Y-filter temperature. Y.-S. Kim et al (2008) reported the development of a CuO NW gas sensor that can detect CO and NO_x in automobiles. The resistance of the semiconductor in semiconductor gas sensor changes when exposed to gases like CO, CO_2, NO_2, etc. The response time and gas sensitivity of semiconductor gas sensors can be improved by using nanosensors with low agglomerated architecture and large surface area\volume ratio. SnO_2, ZnO, and In_2O_3 can also be used as NW materials. Thermal oxidation and solution processing methods can be used for preparing CuO NWs. However, thermal oxidation technique offers advantages such as large aspect

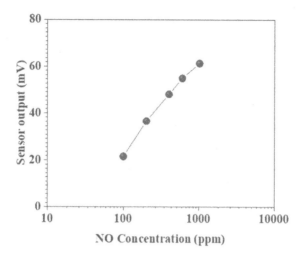

FIGURE 8.1 The zirconia-based sensor output as a function of NO concentration. (S. Zhuiykov, N. Miura, 2007)

ratio and better crystallinity compared with solution processing technique. Let S be the gas response of the sensor which can be computed as (Y.-S. Kim et al, 2008)

$$S = \frac{R_g}{R_a} \qquad (8.2)$$

R_g = resistance of the target gas
R_a = resistance of the pure air.

The sensor response (S) of the CuO NW gas sensor was found to be increased with increase in CO and NO_2 concentration and also with increase in heater power. L. Francioso et al developed an innovative linear temperature gas sensor based on micro hot plate arrays for automotive cabin air quality monitoring that can sense gases like CO, NO_2, and SO_2 (L. Francioso et al, 2008). Perumal Elumalai et al (2009) reported a novel Cr-doped NiO ($Ni_{0.95}Cr_{0.03}O_{1-\delta}$) sensing electrode material prepared using thermal decomposition process for YSZ-based NiO_x sensor. The sensor can detect NO_2 at 800°C under 5 volume % H_2O wet conditions. It is also observed that the conductivity of both NiO and $Ni_{0.95}Cr_{0.03}O_{1-\delta}$ sensing electrodes decreases with increase in temperature. The sensitivity versus NO_2 concentration curves of NiO and $Ni_{0.95}Cr_{0.03}O_{1-\delta}$ sensing electrode-based YSZ-NO_2 sensor under different temperature is shown in Figure 8.2.

The sensitivity increases with increase in NO_2 concentration. Vladimir V. Plashnitsa et al (2009) successfully fabricated a mixed potential type YSZ-planar electro chemical gas sensor. The sensing materials are made of nano structured NiO materials. C. Lopez-Gandara et al (2010) intensively studied the impact of nano structured WO_3 materials on the sensitivity performance of YSZ-based electro chemical NO_x sensor for automotive applications. J.-H Yoon and J.-S. Kim (2011)

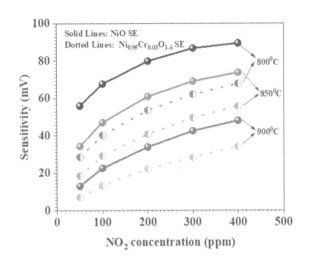

FIGURE 8.2 Sensitivity versus NO_2 concentration curves of NiO and $Ni_{0.95}Cr_{0.03}O_{1-\delta}$ sensing electrode-based YSZ-NO_2 sensor under different temperature. (P. Elumalai et al, 2009)

successfully fabricated a solid-state MEMS type NO_s gas sensor using sol-gel technology. The structure of solid-state MEMS type NO_2 gas sensor is illustrated in Figure 8.3.

It consists of Si-substrate, Pt-sensing electrode, Pt-heating element, SiN_x passivation layer, SiO_2 layer, and SnO_2-WO_3 sensing material. The NO_2 gas sensitivity of solid-state MEMS type NO_2 sensor featuring SnO_2-WO_3 sensing material as a function of atmospheric temperature and NO_2 concentration is shown in Figure 8.4. The MEMS sensor exhibits higher sensitivity at 30°C and higher NO_2 concentration. Heater voltage also affects the sensitivity of the device and it is found that the gas sensitivity increases with increase in heater voltage.

Ying Chen and J.Z. Xiao (2013) reported the development of a potentiometric $La_{1.67}Sr_{0.33}NiO_4$–YSZ NO_x sensor using microwave assisted—complex gel combustion

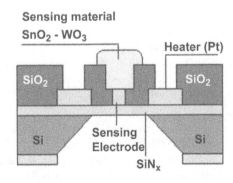

FIGURE 8.3 The structure of solid-state MEMS type NO_2 gas sensor. (J.-H. Yoon, J.-S. Kim, 2011)

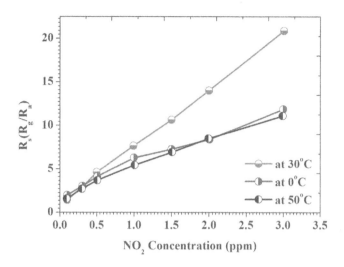

FIGURE 8.4 The NO_2 gas sensitivity of solid-state MEMS type NO_2 sensor featuring SnO_2-WO_3 sensing material as a function of atmospheric temperature and NO_2 concentration. (J.-H. Yoon, J.-S. Kim, 2011)

technique. The sensor structure reported by Ying Chen and J.Z. Xiao (2013) is illustrated in Figure 8.5. The variation of response time as a function of NO concentration for potentiometric $La_{1.67}Sr_{0.33}NiO_4$–YSZ NO_x sensor is depicted in Figure 8.6. The variation of recovery time as a function of NO concentration for potentiometric $La_{1.67}Sr_{0.33}NiO_4$–YSZ NO_x sensor is depicted in Figure 8.7. Sensor with 10 wt% YSZ exhibits faster response time and slower recovery time compared with 20 wt% YSZ and 5 wt% YSZ sensing electrodes.

The influence of Au-YSZ nano-composite electrode on the sensor response characteristics of solid-state potentiometric gas sensor is investigated by (T. Striker et al, 2013). The sensor configuration of Au-YSZ nano-composite electrode based potentiometric gas sensor is shown in Figure 8.8. Todd Striker et al reported that gas sensor with HSA (high surface area) Au-YSZ electrode exhibits higher sensitivity compared with gas sensor with LSA (low surface

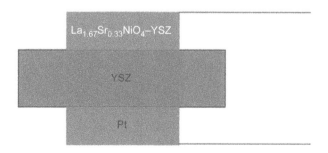

FIGURE 8.5 Potentiometric $La_{1.67}Sr_{0.33}NiO_4$–YSZ NO_x sensor. (Y. Chen, J.Z. Xiao, 2013)

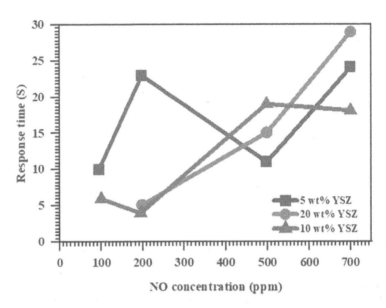

FIGURE 8.6   The variation of response time as a function of NO concentration for potentiometric $La_{1.67}Sr_{0.33}NiO_4$–YSZ NO_x sensor. (Y. Chen, J.Z. Xiao, 2013)

area) Au sensing electrodes. Also the output voltage of the sensor decreases with increase in NO concentration. Yihong Xiao et al (2016) successfully fabricated a solid-state NO_x sensor featuring $GdAlO_3$ perovskite oxide electrolyte. The top view of the $GdAlO_3$ perovskite oxide electrolyte based solid-state NO_x gas sensor is shown in Figure 8.9.

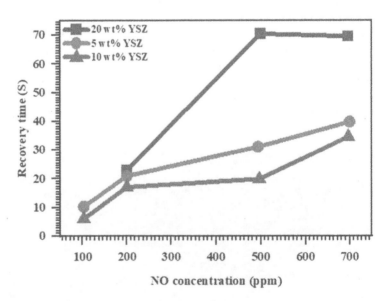

FIGURE 8.7   The variation of recovery time as a function of NO concentration for potentiometric $La_{1.67}Sr_{0.33}NiO_4$–YSZ NO_x sensor. (Y. Chen, J.Z. Xiao, 2013)

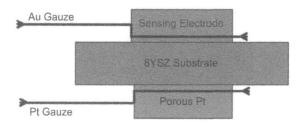

FIGURE 8.8 The sensor configuration of Au-YSZ nano-composite electrode based potentiometric gas sensor. (T. Striker et al, 2013)

It consists of Pt wires, porous NiO layer, GCA $(Gd_{1-x}Ca_xAlO_{3-\delta})$ electrolyte, and a Pt paste layer. The variation of response current as a function of NO_2 concentration for different x values of $Gd_{1-x}Ca_xAlO_{3-\delta}$ substrate is illustrated in Figure 8.10. $Gd_{1-x}Ca_xAlO_{3-\delta}$ substrate with x=0.15 exhibits good response to NO_2 concentration. Increase in Ca results in the increase of conductivity of GCA substrate which leads to higher response current. The tolerance factor (t) of this sensor can be computed as (Y. Xiao et al, 2016)

$$t = \frac{r_A + r_O}{\sqrt{2}\left(r_B + r_O\right)} \tag{8.3}$$

r_A = ionic radii of Gd^+
r_B = ionic radii of Al^{3+}
r_O = ionic radii of O^{2-}.

Fulan Zhong et al (2017) reported that the sensitivity of NO_2 gas sensors can be improved by using alkaline earth metals doped $Gd_2Zr_2O_7$ pyrochlore as O_2 conductors.

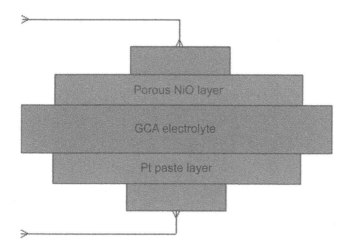

FIGURE 8.9 The top view of the $GdAlO_3$ perovskite oxide electrolyte based solid-state NO_x gas sensor. (Y. Xiao et al, 2016)

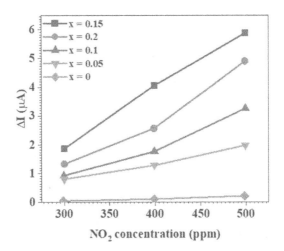

FIGURE 8.10 The variation of response current as a function of NO_2 concentration for different x values of $Gd_{1-x}Ca_xAlO_{3-\delta}$ substrate. (Y. Xiao et al, 2016)

The effect of $Gd_{1.95}Ca_{0.05}Zr_2O_{7+\delta}$, $Gd_{1.95}Ba_{0.05}Zr_2O_{7+\delta}$, and $Gd_{1.95}Sr_{0.05}Zr_2O_{7+\delta}$ O_2 conductors on the sensitivity of NO_2 gas sensors is shown in Figure 8.11. It is found that NO_2 sensor with Ca-doped $Gd_2Zr_2O_7$ O_2 conductor exhibits higher sensitivity compared with Sr and Ba doped $Gd_2Zr_2O_7$ O_2 conductor based NO_2 sensors.

8.4.2 Metal Oxide Semiconductor Nano Gas Sensors

Low manufacturing cost, long life span, less sensitive to temperature and humidity, and high sensitivity are the major advantages of metal oxide semiconductor (MOS)

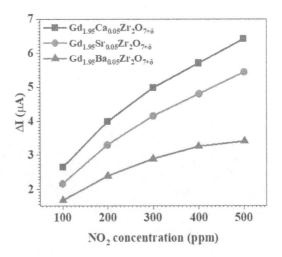

FIGURE 8.11 The effect of $Gd_{1.95}Ca_{0.05}Zr_2O_{7+\delta}$, $Gd_{1.95}Ba_{0.05}Zr_2O_{7+\delta}$, and $Gd_{1.95}Sr_{0.05}Zr_2O_{7+\delta}$ O_2 conductors on the sensitivity of NO_2 gas sensors. (F. Zhong et al, 2017)

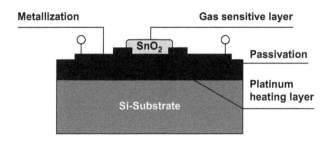

FIGURE 8.12   Si technology based MOS gas sensor. (T. Tille, 2010)

gas sensors. MOS gas sensors works based on the principle of variation in conductivity of a MOS in the presence of an oxidizing or reducing gas (T. Tille, 2010). The structure of a silicon (Si) technology based MOS gas sensor is shown in Figure 8.12. Metal oxide materials such as WO_3, ZnO, SnO_2 are widely used as gas sensing materials. Pt metal is used for metallization and it should be in contact with the metal oxide layer. The metal oxide and metallization layers are isolated from the Pt heating layer by a passivation layer. Reducing gases like C_xH_y and CO causes the increase in conductivity of the MOS whereas the oxidizing gases like NO_2 and O_2 results in the decrease in conductivity of the MOS (T. Tille, 2010).

Good mechanical strength, uniform distribution of temperature in the sensing layer, and low power consumption are the key requirements of a MOS gas sensor. MOS gas sensors can be manufactured in two ways. One is with closed membrane and the second method is with suspended membrane. Silicon, nitrided porous silicon, SiO_2, polysilicon, diamond, SiC, etc., can be used as substrate materials for the production of MOS gas sensors. The heat fluxes of a gas sensor are illustrated in Figure 8.13. Assume the components of heat flow are additive, then the net heat flow can be computed as (I. Simon, 2001)

$$Q_{tot} = G_m \lambda_m \left(T_{hot} - T_{amb} \right) + G_{air} \lambda_{air} \left(T_{hot} - T_{amb} \right) + G_{rad} \sigma \varepsilon \left(T_{hot}^4 - T_{amb}^4 \right) + \Delta x \quad (8.4)$$

where G_{rad}, G_{air}, and G_m represent geometric parameters
T_{hot} = temperature of the hot area
T_{amb} = ambient temperature
λ_m = thermal conductivity of the membrane
λ_{air} = thermal conductivity of the air
ε = emissivity
σ = Stefan-Boltzmann constant.

Micro hot plates can be used to effectively miniaturizing the MOS gas sensors. A micro hot plate consists of a temperature sensor, a thermally isolated area and metal electrodes in contact with the sensing layer. Micro hot plate structure enables high temperature operation at low-power consumption (M. Graf et al, 2004). For achieving uniform temperature distribution across the membrane, symmetric membrane structure is feasible for the micro hot plates. Diego Barrettino et al (2006) reported a monolithic CMOS metal oxide gas sensor for the detection of CO in harsh environments. Due to

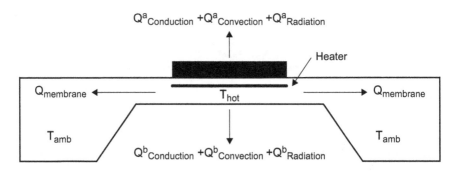

FIGURE 8.13   The heat fluxes of a gas sensor. (I. Simon, 2001)

better reproducibility and excellent stability MEMS, metal oxide gas sensors gained huge attention in the automotive market. The mass production is possible for MEMS metal oxide gas sensors that enables the reduction of cost for sensors. Michael Blaschke et al (2006) reported the development of a MEMS metal oxide gas sensor that can be used for monitoring CO. A micromachined MEMS metal oxide gas sensor consists of Si substrate, Si_3N_4 membrane which acts as the carrier for the sensing layer, Pt heater, and Pt electrodes. MOS gas sensors are attractive due to their ability to directly convert the chemical binding into a voltage signal. The other benefits of MOS gas sensors are they are easy to scale down and low cost production (R.A. Potyrailo et al, 2020). The expose of gases to metal oxides such as SnO_2, ZnO, and tungsten oxide leads to the modification of the electrical resistance. This principle is utilized in sensors.

8.4.3   CO, CO_2, AND C_xH_x NANO GAS SENSORS

H. Okamoto, H. Obayashi, and T. Kudo (1980) reported the development of a stabilized zirconia based CO gas sensor. The response of this sensor is depicted in Figure 8.14.

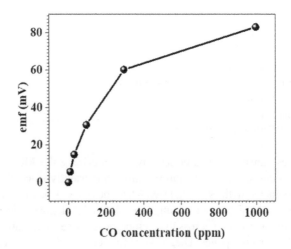

FIGURE 8.14   The response of zirconia-based CO gas sensor. (H. Okamoto, H. Obayashi, T. Kudo, 1980)

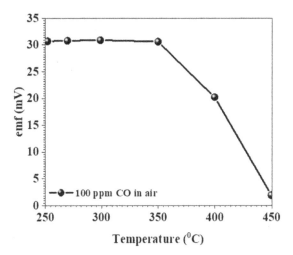

FIGURE 8.15 Temperature-dependent response of zirconia based CO gas sensor. (H. Okamoto, H. Obayashi, T. Kudo, 1980)

V. Demarne and A. Grisel (1988) reported a low-power thin film integrated sensor on Si wafer that can be used to detect CO in automobile exhausts. The temperature also affects the output of stabilized zirconia CO sensor which is demonstrated in Figure 8.15. The structures of mixed potential ceramic sensors that can be used for sensing CO and C_3H_6 in automotive applications are illustrated in Figure 8.16. The response of mixed potential ceramic gas sensor against CO/C_3H_6 gases are demonstrated in Figure 8.17.

A. Dutta, T. Ishihara, and H. Nishiguchi (2004) reported the development of a solid-state amperometric gas sensor for the sensing of C_xH_x in automotive exhaust gas. This sensor features a $LaGaO_3$-based perovskite electrolyte. In_2O_3 and $InSnO_x$, nano powder can also be used as sensing materials for the development of CO gas sensors (G. Neri et al, 2008).

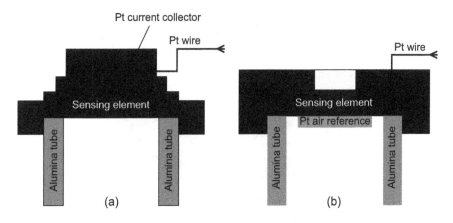

FIGURE 8.16 The structures of mixed potential ceramic sensors that can be used for sensing CO and C_3H_6 in automotive applications. (E.L. Broshaa et al, 2002.)

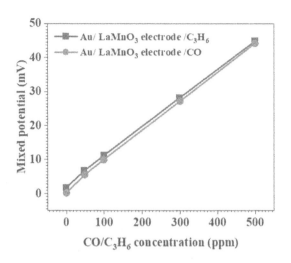

FIGURE 8.17   The response of mixed potential ceramic gas sensor against CO/C_3H_6 gases. (E.L. Broshaa et al, 2002)

8.4.4   HYDROGEN NANO GAS SENSORS

J.S. Suehle, R.E. Cavicchi, and M. Gaitan (1993) reported the first SnO_2 monolithic CMOS micro hot plate gas sensor which can be used for detecting gases like H_2 and O_2 in automotive. SiC FET-based gas sensors are considered as highly suitable for high temperature operation due to the high thermal conductivity of SiC. SiC FET gas sensors exhibits excellent response against H_2 content of the exhaust gas in vehicles. I. Lundström, M.S. Shivaraman, and C.M. Svensson (1975) reported the first-ever Pd-MOS sensor for the detection of H_2 gas. The schematic of a metal insulator SiC (MISiC) sensor that can be used for sensing H_2 is illustrated in Figure 8.18.

MISiC H_2 sensor can be operated over a temperature range of 100–1000°C. Thermal conductivity-based gas sensors are highly attractive for measuring H_2 in automotive applications. The detection of H_2 leakage is very important in vehicles because if H_2 is exposed to ambient air, it will lead to explosion. The structure of a thermal conductivity based micro machined H_2 gas sensor is depicted in Figure 8.19.

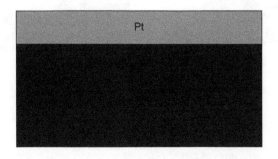

FIGURE 8.18   Cross section of MISiC sensor. (A. Lloyd et al, 1999)

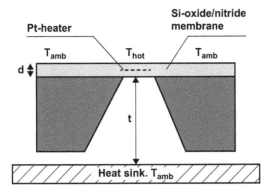

FIGURE 8.19 The structure of a thermal conductivity based micro machined H_2 gas sensor. (I. Simon, M. Arndt, 2002)

Pt and Bulk Si acts as hot element and cold element, respectively. Thermal conductivity based H_2 gas sensing works based on the temperature difference between hot and cold elements. In order to maintain the required temperature difference (ΔT) between hot and cold elements, a power P should be provided which can be computed as (I. Simon, M. Arndt, 2002)

$$P = G_{gas}\lambda_{gas}\Delta T + G_{mem}\lambda_{mem}\Delta T \qquad (8.5)$$

G_{gas} = Geometry of the heat path through gas
λ_{gas} = Thermal conductivity of the gas
G_{mem} = Geometry of the heat path through membrane
λ_{mem} = Thermal conductivity of the membrane.

The efficiency (η) of thermal conductivity-based sensor can be computed as (I. Simon, M. Arndt, 2002)

$$\eta = \frac{G_{gas}\lambda_{gas}}{G_{gas}\lambda_{gas} + G_{mem}\lambda_{mem}}. \qquad (8.6)$$

Thermal time constant (τ) can be used for evaluating the response time of thermal conductivity based H2 gas sensor (I. Simon, M. Arndt, 2002)

$$\tau = R_{therm}C_{therm} \qquad (8.7)$$

R_{therm} = Thermal resistance
C_{therm} = Thermal capacity.

Syed Mubeen et al (2007) reported a SWCNT-based H2 sensor featuring Pd nanoparticles. Pd-decorated CNTs, Pd-meso wires, and Pd-NWs can also be used as sensing materials for the detection of H_2. The resistance change, response time, and

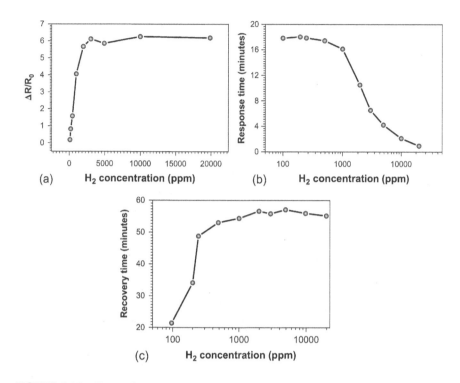

FIGURE 8.20   The resistance change, response time, and recovery time variation of Pd decorated CNT H_2 gas sensor with H_2 concentration. (S. Mubeen et al, 2007).

recovery time variation of Pd decorated CNT H_2 gas sensor with H_2 concentration are depicted in Figure 8.20.

The H_2 sensitivity and recovery time increases with increase in H_2 concentration. Duy-Thach Phan and G.-S. Chung (2014) reported a Pd-graphene nanocomposite-based resistivity type H_2 sensor. The response of Pd-graphene nanocomposite-based resistive type H_2 sensor with H_2 concentration is illustrated in Figure 8.21. The response of Pd-graphene nanocomposite based resistive type H_2 sensor decreases with increase in temperature. H_2 is considered as one of the most promising cleanest energy sources for future automotive vehicles. However, the use of H_2 as fuel poses many safety issues such as rapid and accurate leakage detection to avoid explosion. Acoustic wave sensors, thin film metal sensors and metal oxide sensors can be used for detecting H_2 gas (L. Huang et al, 2015). Juree Hong et al (2015) reported the development of a PMMA polymer membrane coated Pd nanoparticles coated graphene based highly sensitive H_2 gas sensor. N. Lavanya et al (2017) reported the development of a chemically doped SnO_2 based H_2 sensor. Transition metals such as Mn and Co can be used for doping SnO_2. The response of transition metal doped SnO_2 based H_2 sensor is illustrated in Figure 8.22. Xiaohui Tang et al (2019) also reported a Pd-nanocomposite deposited graphene based H_2 sensor for automotive applications. Good H_2 selectivity and superior H_2 solubility are the main advantages of Pd over other metals such as Pt, Au, and Ag.

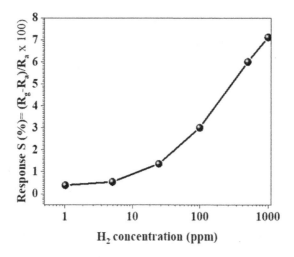

FIGURE 8.21 The response of Pd-graphene nanocomposite based resistive type H_2 sensor with H_2 concentration. (Duy-Thach Phan and G.-S. Chung, 2014)

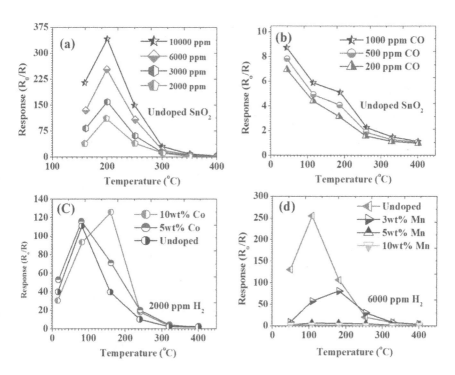

FIGURE 8.22 The response of transition metal doped SnO_2 based H_2 sensor. (N. Lavanya et al, 2017)

8.4.5 OXYGEN NANO GAS SENSORS

Feedback controls of the A/F (air/fuel) ratio-based techniques are popularly used to control emissions and to optimize the efficiency of internal combustion engines. E. M. Logothetis et al (1975) reported a metal oxide A/F oxygen sensor that works based on the principle of change in resistance of metal oxide with the change in ambient O_2 partial pressure. TiO_2 and CoO are the widely used metal oxides for the development of metal oxide A/F oxygen sensors. Zirconia based O_2 sensors can be used to whether the A/F ratio is greater than or less than the stoichiometric ratio (W.J. Fleming 1980). Normally, the electrode takes a porous structure to allow the passage of air into the zirconia surface (R. Usmen, E. Logothetis, M. Shelef, 1995). Planar O_2 sensors are based on planar ZrO_2 technology. Semiconductor metal oxides are used for developing resistive type O_2 sensors. TiO_2 and CoO are mainly used for developing resistive type O_2 sensors. Potentiometric planar ZrO_2 based O_2 sensors works based on Nernst's Law which can be expressed as (E.-I. Tiffée et al, 2001)

$$V = \frac{RT}{4F} \log\left(\frac{P_{O_2}^1}{P_{O_2}^2} \right)$$
(8.8)

R = gas constant
F = Faraday's constant
T = Temperature.

$P_{O_2}^1$ and $P_{O_2}^2$ are the O_2 partial pressure at two different regions. Thimble type ZrO_2 based O_2 sensors are also available to detect O_2 for automotive applications (J. Riegel, H. Neumann, H. Wiedenmann, 2002).

8.5 NANOSENSORS FOR PRESSURE MEASUREMENT

Piezoresistive Si pressure sensors are widely used in automotive applications. Packaging is the major drawback of piezoresistive Si pressure sensors compared with conventional pressure sensors (R.S. Okojie et al, 2015). Silicone-oil filled steel housing is the most popular packaging style used for Si pressure sensors. However, this packaging is costly and consumes large area (H. Krassow, F. Campabadal, E. Lora-Tamayo, 2000) (T.-L. Chou et al, 2009) (M.M. Jevtic et al, 2008). Piezoresistive Si pressure sensors are temperature sensitive and, therefore, it is required to include the thermal compensation techniques in the sensor design. Passive and active temperature compensation methods were reported by Dirk De Bruyker and Robert Puers (2000). K. Birkelund et al (2001) reported a high-pressure Si sensor that can sense pressure ranges from 35 to 1500 bar. Cost effective production, high stability, and long life are the major advantages of Si pressure sensors (K. Birkelund et al, 2001). Capacitive pressure sensors offer high-pressure sensitivity compared with piezoresistive pressure sensors (F. He, Q.-A. Huang, M. Qin, 2007). A Si pressure sensor consists of a thin Si diaphragm that deflects towards the wafer when an external pressure is applied. This deflection of diaphragm induces a capacitance change between the diaphragm and wafer in the case of Si capacitive pressure sensors (V. Tsouti et al, 2007). Liang Lou et al (2011) reported a, Si NW-based piezoresistive pressure sensor. The use of Si NWs enhances the pressure sensitivity of sensors. SOI wafer can be used for reducing the leakage current in Si pressure sensors (X. Lia et al, 2012) (M. Narayanaswamy et al, 2011). Si is an attractive material for making microscale pressure sensors (K. J. Suja et al, 2016) (D. Belavič et al, 1998) (J. Li et al, 2018) (R. Ghosh et al, 2019) (Y. Zhao, Y.-L. Zhao, L.-K. Wang, 2020) (Z. Wanga et al, 2019).

8.5.1 CNT-BASED PRESSURE SENSORS

K. Qian et al (2005) reported a novel CNT array pressure sensor based on field emission concept which has the benefits of fast response, better immune to radiation, low power dissipation, and outstanding thermal stability. The structure of CNT array pressure sensor is shown in Figure 8.23. The fabrication of this sensor involves mainly three steps and they are

> Step 1: Fabrication of anode by wet etching.
> Step 2: MWCNT array growth using hot filament chemical vapor deposition (CVD).
> Step 3: Packaging of the sensor.

Single crystal n-Si wafer acts as anode. 1000-μm thick Si wafer has a resistivity of 0.005 Ω.cm. MWCNT array acts as cathode. The pressure-dependent emission current of CNT array pressure sensor is illustrated in Figure 8.24. The sensor produces high current at high pressure due to the reduction in resistivity.

C. Stampfer et al (2006) reported the fabrication of a SWCNT-based pressure sensor. Narges Doostani et al (2013) also reported a field emission based highly sensitive CNT-based pressure sensor. For the suitable electron emission, it is required to grow

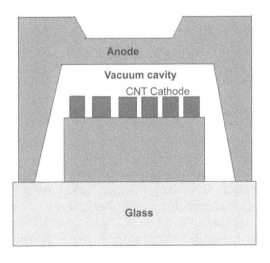

FIGURE 8.23 CNT array pressure sensor. (K. Qian et al, 2005)

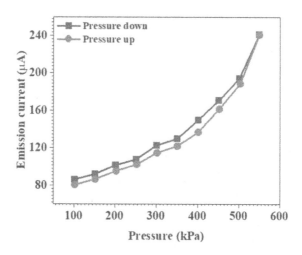

FIGURE 8.24   The pressure-dependent emission current of CNT array pressure sensor. (K. Qian et al, 2005)

vertical CNTs. PECVD process is widely used to grow vertical CNT structures. Pressure sensors can be classified into following three categories:

1. Capacitive pressure sensors.
2. Piezoresistive pressure sensors.
3. Field emission pressure sensors.

Due to low-cost production, small size, and high sensitivity, micromachined pressure sensors are highly desirable in automotive applications. Field emission pressure sensors are advantageous compared with capacitive and piezoresistive pressure sensors. The emission current of the sensor can be computed using Fowler-Nordheim formula which is given below

$$I = AE^2 \exp\left(\frac{-B}{E}\right)$$ (8.9)

where A and B are constants and E represents the electric field applied. The I-V plots of field emission-based CNT pressure sensor at different pressures is depicted in Figure 8.25. The sensor produces higher emission current at high-pressure and high applied voltage between cathode and anode. Anode to emitter distance and anode-cathode potential difference are the two critical factors that influences the emission current (S. Taak et al, 2014) (K. Qian et al, 2005) (K. Qian et al, 2006). L. Wang, D. Lv, and F. Wang (2018) reported a CNT filled silicone rubber composite based piezoresistive pressure sensor (W. Tang et al, 2011). The variation of resistance as a function of pressure for the piezoresistive CNT sensor is shown in Figure 8.26. The resistance of the sensor increases with increase in pressure in sensors with lower CNT content and the resistance of the sensor decreases with increase in pressure in

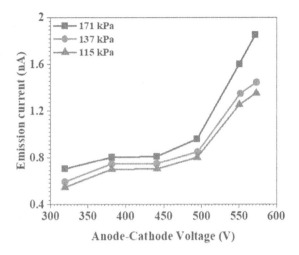

FIGURE 8.25 The I-V plots of field emission based CNT pressure sensor at different pressures. (N. Doostani et al, 2013).

sensors with higher CNT content. CNT FET-based pressure sensors can be used to measure the pressure of automobile tires. Joseph B. Andrews et al (2018) reported a CNT-based thin-film transistor sensor for the measurement of automobile tire pressure. The variation of threshold voltage, transconductance, and on current of CNT-based thin film transistor pressure sensor is illustrated in Figure 8.27.

The variation of resistance of CNT, grapheme, and CNT graphene composite based piezoresistive pressure sensors are illustrated in Figure 8.28. The sensitivity (S) of a piezoresistive pressure sensor can be computed as

$$S = \frac{\Delta R/R_0}{\Delta P}$$ (8.10)

ΔR = DC resistance change
R_0 = Initial resistance at zero pressure
ΔP = External applied pressure difference.

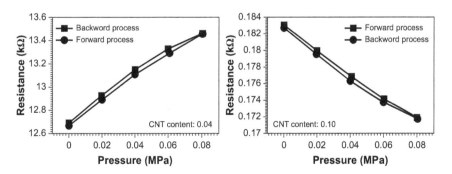

FIGURE 8.26 The variation of resistance as a function of pressure for the piezoresistive CNT sensor. (L. Wang, D. Lv, F. Wang, 2018)

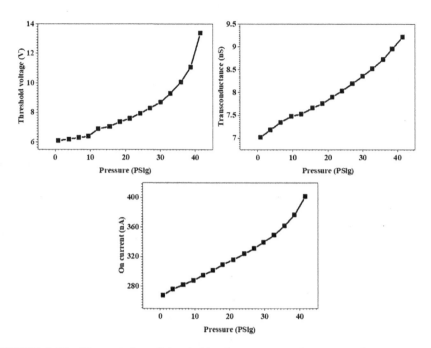

FIGURE 8.27   The variation of threshold voltage, transconductance, and on current of CNT-based thin film transistor pressure sensor. (J.B. Andrews et al, 2018)

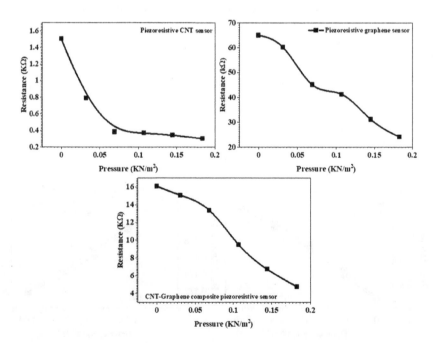

FIGURE 8.28   The variation of resistance of CNT, grapheme, and CNT graphene composite based piezoresistive pressure sensors. (A. Ali et al, 2018)

8.5.2 GRAPHENE PRESSURE SENSORS

Graphene and CNTs are considered as promising materials for the development of pressure nanosensors for the future automotive applications including gas pressure monitoring and tire pressure monitoring (V. Sorkin, Y. W. Zhang, 2011). Shou-En Zhu et al (2013) reported the development of a piezoresistive graphene pressure sensor featuring a SixNy membrane. The variation of output voltage (Vout) with differential pressure of piezoresistive graphene pressure sensor is shown in Figure 8.29. Vout is found to be increasing with increase in differential pressure. M. Habibi et al (2015) reported a field emission-based graphene pressure sensor featuring CNT cathode and reduced graphene oxide sheet anode. The emission current of graphene pressure sensor are illustrated in Figure 8.30. At high pressure, the resonant frequency shift towards the left. J. Aguilera-Servin, T. Miao, and M. Bockrath (2015) and H. Tian et al (2015) also reported a graphene membrane based piezoresistive nano scale pressure sensor. Sung-Ho Shin et al (2017) reported the development of a graphene transistor based pressure sensor for automotive applications which can sense a pressure range of 250 Pa–3 MPa. The sensing characteristic of nanoscale graphene FET pressure sensor is shown in Figure 8.31.

8.5.3 SILICON CARBIDE (SiC) PRESSURE SENSORS

High temperature (200°C–600°C) pressure sensors are essential for automotive applications. The performance of Si pressure sensors severely degrades above 500°C. SiC is considered as one of the most promising materials for high-temperature pressure sensing applications due to their thermal, chemical and electrical stability, high radiation resistance, and good mechanical strength. SiC piezoresistive pressure sensors are highly sensitive to temperature and suffer from variation in contact resistance at high temperatures above 500°C. Therefore, capacitive pressure sensors are

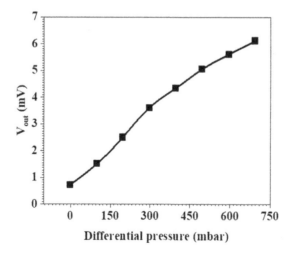

FIGURE 8.29   The variation of output voltage (Vout) with differential pressure of piezoresistive graphene pressure sensor. (S.-E. Zhu et al, 2013)

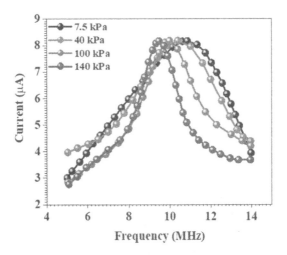

FIGURE 8.30 The emission current of graphene pressure sensor. (M. Habibi et al, 2015).

suitable for high-temperature operation. Darrin J. Young et al (2004) reported the development of a single crystal 3C-SiC high-temperature capacitive sensor for the measurement of pressure in automotive applications. The cross-sectional view of high-temperature capacitive SiC pressure sensor is shown in Figure 8.32. It consists of a Si substrate, sealed cavity, dielectric layer, and a SiC diaphragm. The 3C-SiC diaphragm deflects towards the Si wafer with increase in external pressure which varies the capacitance between wafer and diaphragm. The sensing characteristic of 3C-SiC single crystal capacitive pressure sensor is illustrated in Figure 8.33. C.-H. Wu, C.A. Zorman, and M. Mehregany (2006) reported the development of a high-temperature piezoresistive SiC pressure sensor (Figure 8.34). The defect due to the mismatches in

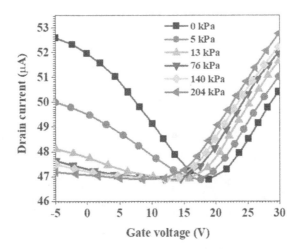

FIGURE 8.31 The sensing characteristic of nanoscale graphene FET pressure sensor. (S.-H, Shin et al, 2017).

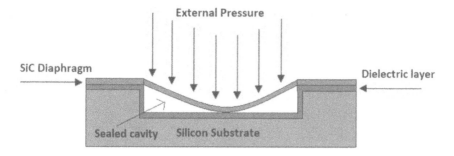

FIGURE 8.32 The cross-sectional view of high-temperature capacitive SiC pressure sensor. (D.J. Young et al, 2004)

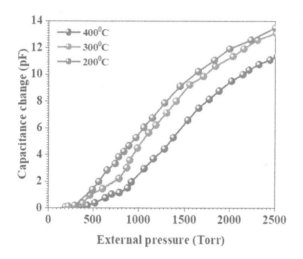

FIGURE 8.33 The sensing characteristics of 3C-SiC single crystal capacitive pressure sensor. (D.J. Young et al, 2004)

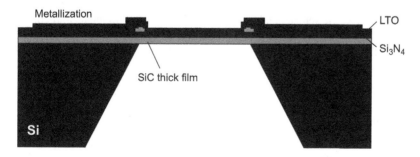

FIGURE 8.34 Bulk micromachined SiC piezoresistive pressure sensor. (C.-H. Wu, C.A. Zorman, and M. Mehregany, 2006)

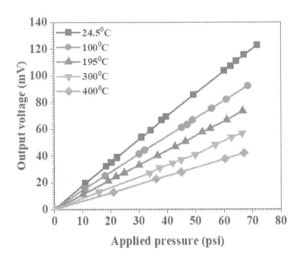

FIGURE 8.35   The variations of output voltage with external applied pressure under different temperatures. (C.-H. Wu, C.A. Zorman, and M. Mehregany, 2006)

lattice constants of Si and SiC layers degrades the sensor performance. The use of SOI wafer helps to minimize the performance degradation. Atmospheric pressure CVD process was used for manufacturing this sensor. The variations of output voltage with external applied pressure under different temperatures are illustrated in Figure 8.35.

SiC capacitive pressure sensors can be used for measuring in-cylinder engine pressure (L. Chen, M. Mehregany 2008). Capacitive SiC pressure sensors can be manufactured using CMOS compatible process (W. Tang et al, 2011). The performance of SiC pressure sensors can be improved by the direct bonding of SiC with treatment of hydrofluoric acid (J. Lia et al, 2020). It is high time to develop high-performance nanosensors and automotive software for enabling power-control toward smart cities (P. Sivakumar et al, 2020) (P. Sivakumar et al, 2016) (R.S. Sandhya Devi, P. Sivakumar, M. Sukanya, 2018).

8.6   CONCLUSION

Nanoscale sensors can be used for monitoring and measuring gases, temperature, and pressure in automotive applications. Silicon-based sensors are suitable for manufacturing microscale sensors and materials such as CNTs, graphene, and compound semiconductors such as GaN, GaAs, InGaAs, and so on are considered as promising materials for developing next-generation nanoscale sensors for automotive applications. The rapid advancements in nanomaterials and nanotechnology also fuel the development of nanoscale sensors for automotive applications.

REFERENCES

J. Aguilera-Servin, T. Miao, M. Bockrath. 2015. Nanoscale pressure sensors realized from suspended graphene membrane devices, Applied Physics Letters, 106, 083103. doi: 10.1063/1.4908176.

A. Ali, A. Khan, Kh.S. Karimov, A. Ali, A.D. Khan. 2018. Pressure sensitive sensors based on carbon nanotubes, graphene, and its composites, Hindawi Journal Nanomaterials, 2018, 9592610. https://doi.org/10.1155/2018/9592610.

J.B. Andrews, J.A. Cardenas, C.J. Lim, S.G. Noyce, J. Mullett, A.D. Franklin. 2018. Fully Printed and Flexible Carbon Nanotube Transistors for Pressure Sensing In Automobile Tires, IEEE Sensors Journal, 18(19), 7875–7880.

D. Barrettino, M. Graf, S. Taschini, S. Hafizovic, C. Hagleitner, A. Hierlemann. 2006. CMOS monolithic metal–oxide gas sensor microsystems, IEEE Sensors Journal, 6(2), 276–286.

D. Belavič, Š. Stojan, M. Pavlin, D. Ročak, M. Hrovat. (1998). Silicon pressure sensors with a thick film periphery, Microelectronics International, 15(3), 26–30.

E. Billi, J.-P. Viricelle, L. Montanaro, C. Pijolat. 2002. Development of a protected gas sensor for exhaust automotive applications, IEEE Sensors Journal, 2(4), 342–348.

K. Birkelund, P. Gravesen, S. Shiryaev, P.B. Rasmussen, M.D. Rasmussen. 2001, High-pressure silicon sensor with low-cost packaging, Sensors and Actuators A: Physical, 92(1–3), 16–22.

M. Blaschke, T. Tille, P. Robertson, S. Mair, U. Weimar, H. Ulmer. 2006. MEMS gas-sensor array for monitoring the perceived car-cabin air quality, IEEE Sensors Journal, 6(5), 1298–308.

E.L. Broshaa, R. Mukundan, D.R. Brown, F.H. Garzon, J.H. Visser. 2002. Development of ceramic mixed potential sensors for automotive applications, Solid State Ionics, 148 (1–2), 61–69.

L. Chen, M. Mehregany. 2008. A silicon carbide capacitive pressure sensor for in-cylinder pressure measurement, Sensors and Actuators A: Physical, 145, 2–8.

Y. Chen, J.Z. Xiao. 2013. Synthesis of composite $La_{1.67}Sr_{0.33}NiO_4$–YSZ for a potentiometric NOx sensor by microwave-assisted complex-gel auto-combustion, Ceramics International, 39 (8), 9599–9603.

T.-L. Chou, C.-H. Chu, C.-T. Lin, K.-N. Chiang. 2009. Sensitivity analysis of packaging effect of silicon-based piezoresistive pressure sensor, Sensors and Actuators A, 152(1), 29–38.

V. Demarne, A. Grisel. 1988. An integrated low-power thin-film co gas sensor on silicon, Sensor and Actuators, 13(4), 301–313.

D. De Bruyker, R. Puers. 2000. Thermostatic control for temperature compensation of a silicon pressure sensor, Sensors and Actuators 82, 120–127.

N. Doostani, S. Darbari, S. Mohajerzadeh, M.K. M-Farshi. 2013, Fabrication of highly sensitive field emission based pressure sensor, using CNTs grown on micro-machined substrate, Sensors and Actuators A: Physical, 201, 310–315.

A. Dutta, T. Ishihara, H. Nishiguchi. 2004. An amperometric solid-state gas sensor using aLaGaO3-based perovskite oxide electrolyte for detecting hydrocarbon in exhaust gas. A bimetallic anode for improving sensitivity at low temperature, Chemistry of Materials, 16, 5198–5204.

P. Elumalai, J. Zosel, U. Guth, N. Miura. 2009. NO2 sensing properties of YSZ-based sensor using NiO and Cr-doped NiO sensing electrodes at high temperature, Ionics 15, 405–411.

W.J. Fleming. 1980. Zirconia oxygen sensor–An equivalent circuit model, SAE Transactions, 89, 76–90. doi.org/10.4271/800020.

L. Francioso, A. Forleo, A. Taurino, P. Siciliano, L. Lorenzelli, V. Guarnieri, et al. 2008. Linear temperature microhotplate gas sensor array for automotive cabin air quality monitoring, Sensors and Actuators B: Chemical, 134(2), 660–665.

S. Garrigues, T. Talou, D. Nesa. 2004. Comparative study between gas sensors arrays device, sensory evaluation and GC/MS analysis for QC in automotive industry, Sensors and Actuators B: Chemical, 103(1–2) 55–68.

R. Ghosh, et al. 2019. High resolution wide range pressure sensor using hexagonal ring and micromachined cantilever tips on 2D silicon photonic crystal, Optics Communications, 431, 93–100. https://doi.org/10.1016/j.optcom.2018.09.016.

M. Graf, D. Barrettino, S. Taschini, C. Hagleitner, A. Hierlemann, H. Baltes. 2004. Metal oxide-based monolithic complementary metal oxide semiconductor gas sensor microsystem, Analytical Chemistry, 76(15), 4437–4445.

N. Guillet, R. Lalauze, J.-P. Viricelle, C. Pijolat, L. Montanaro. 2002. Development of a gas sensor by thick film technology for automotive applications: Choice of materials—realization of a prototype, Materials Science and Engineering: C, 21(1–2), 97–103.

M. Habibi, S. Darbari, S. Rajabali, V. Ahmadi. 2015. Fabrication of a graphene-based pressure sensor, taking advantage of field emission behavior of carbon nanotubes, Elseiver 96, 259–267. doi: 10.1016/j.carbon.2015.09.059.

G. Hagen, A. Harsch, R. Moos. 2018. A pathway to eliminate the gas flow dependency of a hydrocarbon sensor for automotive exhaust applications, Journal of Sensors and Sensor Systems, 7, 79–84.

F. He, Q.-A. Huang, M. Qin. 2007. A silicon directly bonded capacitive absolute pressure sensor, Sensors & Actuators: A. Physical, 135, 507–514.

J. Hong, S. Lee, J. Seo, S. Pyo, J. Kim, T. Lee. 2015. A highly sensitive hydrogen sensor with gas selectivity using a PMMA membrane-coated Pd nanoparticle/single-layer graphene hybrid, ACS Applied Materials Interfaces, 7, 3554–3561.

L. Huang, Z. Zhang, Z. Li, B. Chen, X. Ma, L. Dong, L.-M. Peng. 2015. Multifunctional graphene sensors for magnetic and hydrogen detection, ACS Applied Materials Interfaces, 7(18), 9581–9588.

M.M. Jevtic et al. 2008. Diagnostic of silicon piezoresistive pressure sensors by low frequency noise measurements, Sensors and Actuators A, 144, 267–274.

Y.-S. Kim, I.-S. Hwang, S.-J. Kim, C.-Y. Lee, J.-H. Lee. 2008. CuO nanowire gas sensors for air quality control in automotive cabin, Sensors and Actuators B: Chemical, 135(1), 298–303.

H. Krassow, F. Campabadal, E. Lora-Tamayo. 2000. Wafer level packaging of silicon pressure sensors, Sensors and Actuators, 82 (1–3), 229–233.

N. Lavanya, C. Sekar, E. Fazio, F. Neri, S. Leonardi. 2017. Development of a selective hydrogen leak sensor based on chemically doped SnO2 for automotive applications, International Journal of Hydrogen Energy, 42, 10645–10655.

J. Li, D. Geng, D. Zhang, W. Qin, Y. Jiang. 2018. Ultrasonic vibration mill-grinding of single-crystal silicon carbide for pressure sensor diaphragms, Ceramics International, 44(30), 3107–3112.

J. Lia, Y. Jianga, H. Lia, X. Liangc, M. Huang, D. Liuc, D. Zhanga. 2020. Direct bonding of silicon carbide with hydrofluoric acid treatment for high temperature pressure sensors, Ceramics International, 46(3), 3944–3948.

X. Lia et al. 2012. High-temperature piezoresistive pressure sensor based on implantation of oxygen into silicon wafer, Sensors and Actuators A, 179, 277–282.

A. Lloyd, P. Tobias, A. Baranzahi, P. Martensson, I. Lundstrom. 1999. Current status of silicon carbide based high-temperature gas sensors, IEEE Transactions on Electron Devices, 46(3) 561–566.

E. Logothetis. 1975. Metal oxide oxygen sensors for automotive applications, Journal of Solid State Chemistry France, 12(3–4), 331.

C. López-Gándara, J.M. Fernández-Sanjuán, F.M. Ramos, A. Cirera. 2010. Effect of nanostructured WO3 layers in the sensitivity to nitrogen oxide in YSZ-based electrochemical sensors for automotive applications, Procedia Engineering, 5, 164–167.

L. Lou, S. Zhang, L. Lim, W.-T. Park, H. Feng, D.-L. Kwong, C. Lee. 2011. Characteristics of NEMS piezoresistive silicon nanowires pressure sensors with various diaphragm layers, Procedia Engineering, 25, 1433–1436.

I. Lundström, M.S. Shivaraman, C.M. Svensson. 1975. Hydrogen sensitive MOS field effect transistor, Applied Physics Letters, 26(2), 55–57.

S. Mubeen, T. Zhang, B. Yoo, M.A. Deshusses, N.V. Myung. 2007. Palladium nanoparticles decorated single-walled carbon nanotube hydrogen sensor, Journal of Physical Chemistry C, 111, 6321–6327.

M. Narayanaswamy, R. Joseph Daniel, K. Sumangala, C.A. Jeyasehar. 2011. Computer aided modelling and diaphragm design approach for high sensitivity silicon-on-insulator pressure sensors, Measurement, 44(10), 1924–1936.

G. Neri, A. Bonavita, G. Micali, G. Rizzo, E. Callone, G. Carturan. 2008. Resistive CO gas sensors based on In2O3 and InSnOx nanopowders synthesized via starch-aided sol–gel process for automotive applications, Sensors and Actuators B: Chemical, 132(1), 224–233.

H. Okamoto, H. Obayashi, T. Kudo. 1980. Carbon monoxide gas sensor made of stabilized zirconia, Solid State Ionics, 1(3–4), 319–326.

R.S. Okojie, D. Lukco, V. Nguyen, E. Savrun 2015. 4H-SiC piezoresistive pressure sensors at 800°C with observed sensitivity recovery, IEEE Electron Device Letters, 36(2) 174–176.

D.-T. Phan, G.-S. Chung 2014. Characteristics of resistivity-type hydrogen sensing based on palladium-graphene nanocomposites, International Journal of Hydrogen Energy, 39, 620–629.

V.V. Plashnitsa, P. Elumalaia, Y. Fujioc, N. Miura. 2009. Zirconia-based electrochemical gas sensors using nano-structured sensing materials aiming at detection of automotive exhausts, Electrochimica Acta 54, 6099–6106.

R.A. Potyrailo, S. Go, D. Sexton et al. 2020. Extraordinary performance of semiconducting metal oxide gas sensors using dielectric excitation, Nature Electronics, 3, 280–289.

K. Qian et al. 2006. Studies on vacuum microelectronic pressure sensors based on carbon nanotubes arrays, Physica E, Low-Dimensional Systems Nanostructure, 31(1), 1–4.

K. Qian, T. Chen, B. Yan, Y. Lin, D. Xu, Z. Sun, B. Cai 2005. Research on carbon nanotube array field emission pressure sensors, Electronics Letters, 41, 824–825.

J. Riegel, H. Neumann, H. Wiedenmann. 2002. Exhaust gas sensors for automotive emission control, Solid State Ionics, 152–153, 783–800.

T. Ritter, G. Hagen, J. Lattus, R. Moos. 2018. Solid state mixed-potential sensors as direct conversion sensors for automotive catalysts, Sensors and Actuators B: Chemical, 255(3), 3025–3032.

R.S. Sandhya Devi, P. Sivakumar, M. Sukanya. 2018. Offline analysis of sensor can protocol logs without can/vector tool usage, International Journal of Innovative Technology and Exploring Engineering 8 pp. 225–229.

S. Saponara, E. Petri, L. Fanucci, P. Terreni. 2011. Sensor modeling, low-complexity fusion algorithms, and mixed-signal IC prototyping for gas measures in low-emission vehicles, IEEE Transactions on Instrumentation and Measurement, 60(2), 372–384.

S.-H. Shin, et al. 2017. Integrated arrays of air-dielectric graphene transistors as transparent active-matrix pressure sensors for wide pressure ranges. Nature Communications, 8, 14950.

I. Simon, M. Arndt. 2002. Thermal and gas-sensing properties of a micromachined thermal conductivity sensor for the detection of hydrogen in automotive applications, Sensors and Actuators A: Physical, 97, 104–108.

I. Simon, N.B. Ãrsan, M. Bauer, U. Weimar. 2001. Micromachined metal oxide gas sensors: Opportunities to improve sensor performance, Sensors and Actuators B, 73(1), 1–26.

P. Sivakumar, R. Nagaraju, D. Samanta, M. Sivaram, M.N. Hindia, I.S. Amiri. 2020. A novel free space communication system using nonlinear InGaAsP microsystem resonators for enabling power-control toward smart cities, Wireless Networks, 26, 2317–2328.

P. Sivakumar, B. Vinod, R.S. Sandhya Devi, R. Divya. 2016. Deployment of effective testing methodology in automotive software development, Circuits and Systems, 7(9), 2568–2577.

V. Sorkin, Y.W. Zhang. 2011. Graphene-based pressure nanosensors, Journal of Molecular Modeling, 17, 2825–2830. doi 10.1007/s00894-011-0972-0.

C. Stampfer, T. Helbling, D. Obergfell, B.S. berle, M.K. Tripp, A. Jungen, S. Roth, V.M. Bright, C. Hierold. 2006. Fabrication of single-walled carbon-nanotube-based pressure sensors, Nano Letters, 6(2), 233–237.

T. Striker, V. Ramaswamy, E.N. Armstrong, P.D. Willson, E.D. Wachsman, J.A. Ruud. 2013. Effect of nanocomposite Au–YSZ electrodes on potentiometric sensor response to NOx and CO, Sensors and Actuators B: Chemical, 181, 312–318.

J.S. Suehle, R.E. Cavicchi, M. Gaitan. 1993. Tin oxide gas sensor fabricated using CMOS micro-hotplates and in -situ processing, IEEE Electron Device Letters, 14, 118–120.

K.J. Suja, G.S. Kumar, R. Komaragiri, A. Nisanth 2016. Analysing the effects of temperature and doping concentration in silicon based MEMS piezoresistive pressure sensor, Procedia Computer Science, 93, 108–116.

S. Taak, S. Rajabali, S. Darbari, S. Mohajerzadeh. 2014. High sensitive/wide dynamic range, field emission pressure sensor based on fully embedded CNTs, Journal of Physics D, 47, 045302.

W. Tang et al. 2011. Complementary metal-oxide semiconductor-compatible silicon carbide pressure sensors based on bulk micromachining, Micro & Nano Letters, 6(4), 265.

X. Tang, P. Haddad, N. Mager et al. 2019. Chemically deposited palladium nanoparticles on graphene for hydrogen sensor applications, Scientific Reports 9, 3653.

H. Tian et al. 2015. A graphene-based resistive pressure sensor with record-high sensitivity in a wide pressure range, Science. Rep. 5, 8603.

E.-I. Tiffée, K. Härdtl, W. Menesklou, J. Riegel. 2001. Principles of solid state oxygen sensors for lean combustion gas control, Electrochimica Acta, 47, 807–14.

T. Tille 2010. Automotive requirements for sensors using air quality gas sensors as an example, Procedia Engineering, 5, 5–8.

V. Tsouti, G. Bikakis, S. Chatzandroulis, D. Goustouridis, P. Normand. 2007. Impact of structural parameters on the performance of silicon micromachined capacitive pressure sensors, Sensors and Actuators A: Physical, 137 (1), 20–24.

R. Usmen, E. Logothetis, M. Shelef. 1995. Measurement of Pt electrode surface area of automotive ZrO_2 oxygen sensors, Sensors and Actuators B: Chemical, 28(2), 139–42.

L. Wang, D. Lv, F. Wang. 2018. Electrode-shared differential configuration for pressure sensor made of carbon nanotube-filled silicone rubber composites, IEEE Transactions on Instrumentation and Measurement, 67(6), 1417–1424.

Z. Wanga, B. Sia, S. Chena, B. Jiao, X. Yan. 2019. A nondestructive Raman spectra stress 2D analysis for the pressure sensor sensitive silicon membrane, Engineering Failure Analysis, 105, 1252–1261.

D.L. West, F.C. Montgomery, T.R. Armstrong 2005. "NO-selective" NOx sensing elements for combustion exhausts, Sensors and Actuators B: Chemical, 111, 84–90.

C.-H. Wu, C.A. Zorman, M. Mehregany. 2006. Fabrication and testing of bulk micromachined silicon carbide piezoresistive pressure sensors for high temperature applications, IEEE Sensors, 6(2), 316–24.

Y. Xiao, D. Wang, G. Cai, Y. Zheng, F. Zhong. 2016. A $GdAlO_3$ perovskite oxide electrolyte-based NOx solid-state sensor, Science Report, 6, 37795. doi: 10.1038/srep37795.

J.C. Yang, P.K. Dutta. 2007. Promoting selectivity and sensitivity for a high temperature YSZ-based electrochemical total NOx sensor by using a Pt-loaded zeolite Y filter, Sensors and Actuators B: Chemical, 125(1), 30–39.

J.-H. Yoon, J.-S. Kim. 2011. Study on the MEMS-type gas sensor for detecting a nitrogen oxide gas, Solid State Ionics, 192(1), 668–671.

D.J. Young, J. Du, C.A. Zorman, W.H. Ko. 2004, High-temperature single-crystal 3C-SiC capacitive pressure sensor, IEEE Sensors Journal, 4(4), 464–470.

H. Zhang, J. Wang, Y.-Y. Wang. 2015. Sensor reduction in diesel engine two-cell selective catalytic reduction (SCR) systems for automotive applications, IEEE/ASME Transactions on Mechatronics, 20(5), 2222–2233.

Y. Zhao, Y.-L. Zhao, L.-K. Wang. 2020. Application of femtosecond laser micromachining in silicon carbide deep etching for fabricating sensitive diaphragm of high temperature pressure sensor, Sensors and Actuators A: Physical, 309, 112017.

F. Zhong, J. Zhao, L. Shi et al. 2017. Alkaline-earth metals-doped pyrochlore $Gd_2Zr_2O_7$ as oxygen conductors for improved NO_2 sensing performance. Science Report, 7, 4684.

S.-E. Zhu, M.K. Ghatkesar, C. Zhang, G.C.A.M. Janssen. 2013. Graphene based piezoresistive pressure sensor, Applied Physics Letters, 102, 161904. doi:10.1063/1.4802799.

S. Zhuiykov, N. Miura. 2007. Development of zirconia-based potentiometric NOx sensors for automotive and energy industries in the early 21st century: What are the prospects for sensors? Sensors and Actuators B: Chemical, 121(2), 639–651.

Index